LIVING WITH THUNDER

Living with Thunder

Exploring the Geologic Past, Present, and Future of the Pacific Northwest

ELLEN MORRIS BISHOP

OREGON STATE UNIVERSITY PRESS   CORVALLIS

The John and Shirley Byrne Fund for Books on Nature and
the Environment provides generous support that helps make
publication of this and other Oregon State University Press
books possible.

The paper in this book meets the guidelines for permanence
and durability of the Committee on Production Guidelines
for Book Longevity of the Council on Library Resources
and the minimum requirements of the American National
Standard for Permanence of Paper for Printed Library
Materials Z39.48-1984.

Library of Congress Cataloging-in-Publication Data

Bishop, Ellen Morris, author.
  Living with thunder : exploring the geologic past, present,
and future of the Pacific Northwest / Ellen Morris Bishop.
    pages cm
  Includes bibliographical references.
  ISBN 978-0-87071-748-2 (pbk. : alk. paper)
  1. Geology—Northwest, Pacific. I. Title.
  QE79.B57 2014
  557.97—dc23

                  2014019337

First published in 2014 by Oregon State University Press
Printed in China

Oregon State University Press
121 The Valley Library
Corvallis OR 97331-4501
541-737-3166 • fax 541-737-3170
www.osupress.oregonstate.edu

FOR DAVE:

Time

Is but an instant

Compared to

The endurance of

Love.

ACKNOWLEDGMENTS

This book would never have been written without the patience and support of my husband, David Bishop, whose inspiration and encouragement were essential to its completion. Thanks also to my family and friends, especially Beth Bishop, Jan van Orsow, Susan Goolsbee, and Cory Blackeagle for helping in many ways at many times. Manuscript reviews by Marli Miller, David Montgomery, and an anonymous reviewer greatly improved the flow and content of the book. Amy Molitor designed and produced the geographical maps; Mollie Firestone's editorial expertise helped greatly to perfect the manuscript.

CONTENTS

Crater Lake and Wizard Island, looking east toward Mount Scott on far side. Crater Lake was created by the cataclysmic eruption of Mount Mazama about 7,700 years ago.

INTRODUCTION The Volcanic Heritage of the Pacific Northwest

The Pacific Northwest's landscape celebrates volcanoes. Visit any Northwest icon—Mount Rainier, Haystack Rock, Multnomah Falls, Hells Canyon, or even the Owyhees' Leslie Gulch—and geologists will likely identify a volcano as its progenitor. Perhaps the most obvious example is Crater Lake, whose thunderous inception was recorded in oral histories by those who witnessed it. The Klamath Indians, a people who observed the cataclysmic eruption of Moy-Yaina (Mount Mazama) 7,700 years ago, explain the volcano's behavior this way on the tribal website:

> The underworld chief, Monadalkni, sought to marry the beautiful maiden Loha, daughter of the Klamath's chief. Rejected by the young woman, Monadalkni returned to his dark kingdom and shook the mountain in rage, thundering loudly as he stormed through the underworld passages, and hurling lightning bolts towards his would-be lover on the surface. The thunder caused the top of Moy-Yaina (Mount Mazama) to explode. Great fireballs shot out, accompanied with five great booms of thunder. Ultimately, Gmok'am'c, The Creator, battled with the evil Monadalkni, who continued to bellow with thunderous roars until Gmok'am'c forced him back underground, and sealed the fuming entrance to the underworld with hot rocks, sulfur, and ash. Then Gmok'am'c made rain to fill the remaining hole with water, transforming it into a beautiful and sacred place. The lake is called Giiwaas—most sacred place. The truncated mountain is Tum-sum-ne.

We are no strangers to Monadalkni's voice. The Northwest is the offspring of volcanoes, from 700 million years ago to the present. Monadalkni's thunderous roar has accompanied the creation of offshore volcanic islands and basaltic seafloor, the production of the "granites" of the Wallowas, North Cascades, Klamaths, and Sierra. Some 60 to 50 million years ago, Monadalkni's fluid rage emerged as the basaltic lavas of submarine volcanoes not far offshore. His wrath heralded the soft red rhyolites and the explosive volcanoes of the San Poil Valley 50 million years ago, and the younger Mutton Mountains, Ochocos, and the High Desert. The dark basalts of the Columbia Plateau arrived with a prolonged rumble rather than a roar. The Pliocene Cascades, fraught with explosive eruptions 5.33 to 2.59 million years ago, would have howled. The modern Cascades raged against Pleistocene ice. Today the Cascades still shudder as Monadalkni paces his chambers.

No other field of science is as rich in tales as geology. We geologists are story-weavers as much as the Klamath people, who, with acuity born of firsthand experience, provided a detailed portrait of Mazama's catastrophic eruption.

However, only the stones themselves can bear witness to the planet's more ancient past. But stones are mute. It is geology's charge to give voice to the planet. To make sense of all our data, we, like the Klamaths, must tell the story. This lava flowed from here to there. It engulfed an aged rhino on its way, and flowed through a forest of bald cypress, metasequoia, spruce, through a small valley and a

Rainier Sunrise: Mount Rainier, Tahoma Peak (left), and Emmons Glacier at dawn, from the Sunrise area. Mount Rainier ranks as one of the Northwest's most likely candidates for noteworthy and destructive future activity.

swamp, into a bigger river—see, look, here is where the lava encountered the stream, and here where it made a dam. Did salmon leap over the barrier? Was there a lake? Did the other rhinos escape? Imagination resides just beneath the surface in science, and geology is perhaps the most thin-skinned of all.

Our intimacy with stones runs deeper than landscape. For better or worse, we are literally part of the planet, composed not only of dust, but of earth, stars, and former life. Whatever we are made of has been worn, used, eaten, grown, erupted, pooped, upchucked, and recycled countless times before our individual use. The oxygen molecules of the Permian air, the carbon held briefly in the long, tropical leaves of Eocene forests, the iron that inhabited a torrid basaltic lava, these are now part of us. The phosphorous in our front tooth may have once built the tooth of a tyrannosaur or the tusk of a woolly mammoth. The oxygen that fuels our cells may be the same atom that once journeyed through the lungs of a triceratops. It might have been cycled into the atmosphere 2.4 billion years ago by blue-green algae in the first riotous oxygen orgy of photosynthesis. It might have been inhaled by Secretariat on the home stretch to winning the Triple Crown, or exhaled by the tallest of redwoods. We are the stuff of the planet. Its history is literally in our flesh and blood and bones.

Indigenous peoples have long understood this blood relation between the Earth and people. There is, for example, the Nimi'ipuu (Nez Perce) explanation of their creation. Once, so the story goes, It'se-ye-ye (Coyote) was fixing Celilo Falls so that Salmon could negotiate this barrier more easily. Fox hurried up to him and asked if Coyote could do something to save the animal people from a monster of huge and unmeasured proportions that had gobbled up everyone, and lay in the Clearwater River valley, near Kamiah, Idaho. Coyote went upriver and was inhaled by the monster, just like everyone else. But crafty Coyote had brought his stone knives and flints with him and, once inside the monster, lit a fire beneath the monster's heart, and then cut it loose with his stone knives. Everyone was saved. Coyote, the Trickster, sliced up the monster's heart and flung the pieces in four directions. All the other tribes sprang up from these fragments. Fox reminded Coyote that there were still no humans at Kamiah. So the Trickster dipped his hands into the remainder of the monster's heart and shook the drops of blood onto the ground. The Nimi'ipuu formed from these. As he shook his hands, Coyote said "*q'o' 'óykalapa cicíkaw'is, hahamaníx, kaa titác̓ titóoqan 'éetx pewc̓éeyu'.*" "You will be small people, but you will, in everything, be brave and good people."

Today, geologists recognize the bulbous outcropping of the Heart of the Monster as a Columbia River basalt dike—the source of one of the earliest of the stacked flows that formed the Nimi'ipuu's home landscapes in the canyons and plateaus of the Snake and Clearwater Rivers. To choose one of the greatest of volcanic flows as the place of origin

Near Kamiah, Idaho, a basalt outcropping is the Heart of the Monster. The Nimi'ipuu (Nez Perce) people arose from this place

may seem cultural and geologic serendipity to some. But it also speaks of a visceral understanding, bone-deep and rooted in psyche, of a connection to landscape.

So we—geologists, indigenous peoples, and, in fact, all living things—are intimate parts of the planet and its history. Geology lives in us all, along with the syncopated rhythms of time and the inhalations of all those creatures who thrived long before our coming. To understand geology, and geologic history, is to better understand ourselves.

Steens Mountain in southeast Oregon was the site of multiple eruptions from 21 to 16.5 million years ago. Its 9,733-foot summit was lifted more than a mile above the adjacent Alvord Desert by faulting.

CHAPTER 1   Pacific Northwest Geology   *The Big Picture*

The Pacific Northwest is a region defined as much by its geology as by its drippy or dry landscapes, its threatened salmon, and its inhabitants' proclivity for a green lifestyle. Geographers define it as the region influenced by Pacific Maritime weather, the Columbia River, and Cascade volcanoes. For the most parochial, the Pacific Northwest encompasses only Oregon and Washington. Maybe Idaho. A snippet of northern California's Cascades and Klamaths and perhaps the Modoc Plateau. Geographers toss in British Columbia and some scraps of Alberta, to the foot of the Rockies.

For geologists, it is the subduction zones and accreted terranes, flood basalts, bombastic volcanoes, and a smattering of native sediments that provide the essential character of the place. We are Mount St. Helens. We are the Columbia River Gorge. We are fitful volcanoes, unrepentant basalts, and a collage of ancient volcanic islands, seamounts, and ocean bottoms.

This book is intended as an introductory history of the Pacific Northwest—Washington, Oregon, northern California, and western Idaho—for the general reader and geologic nonspecialist. It begins in the Archean world, 2.5 billion years ago, when the most sophisticated life-forms were algae and cyanobacteria and the Earth's atmosphere was bereft of oxygen. The rocks that represent this time include the Priest River complex of northwest Washington and western Idaho—volcanic rocks and intrusives (magmas that do not breach the surface) that formed part of the early North American crust.

By a billion years ago, the place that would become the Northwest was a shallow sea floored by sands from North America to the east, and from a highland that would become Australia to the west. The Selkirk Mountains in northeastern Washington bear evidence of a long global ice age, Snowball Earth, which lasted from 780 million years to 630 million years ago. Then, about 600 million years ago,

as volcanic eruptions warmed the globe and ice retreated, the planet's first complex life appeared. Known as Ediacarans, these small, mysterious marine creatures are preserved as diaphanous fossils in the Klamath Mountains, near Yreka, California.

By 540 million years, trilobites—tiny, three-lobed marine animals that bear a resemblance to modern sowbugs—appear in the limestones near Metaline Falls, Washington. Other denizens of the Cambrian seas included sponges and early attempts at corals.

Between 400 million and 150 million years ago, volcanic archipelagoes and microcontinents prospered in the sea west of the Idaho and eastern Washington shore. As the Atlantic Ocean opened, and North America moved westward, islands in the cluttered seas collided with the continent, building its shoreline westward. The last of these collisions occurred about 50 million years ago, adding the Oregon Coast Range, Willapa Hills, and Olympic Mountains.

Since that time, volcanic activity has built and buttressed the landscape, including the eruption of supervolcanoes 40 to 28 million years ago, and the appearance of Columbia River and Steens flood basalts 16.7 million to 5.5 million years ago. Today, these define the landscapes of eastern Oregon and eastern Washington, as well as coastal Oregon. Beginning about 15 million years ago, the crust of southeastern Oregon and Nevada began stretching, producing the Basin and Range, and also uplifting the Blue Mountains and the ridges of central Washington.

The Pleistocene (Ice Age), 2.59 million to only 11,700 years ago, remodeled our landscape using ice and volcanic fire. Today, we inhabit an active time that includes fitful Cascade volcanoes, a querulous subduction zone, and unpredictable faults that owe allegiance to the expanding Basin and Range.

In addition, we face the challenges of a warming globe and changing climate. In this, geologic history offers much experience but little solace. Global and Northwest records demonstrate that with rapid climate change comes ecosystem malaise and, often, calamitous extinctions.

In the Northwest, the times with the strongest resonance and greatest representation are found in the Cenozoic Era—the geologic periods of the past 66 million years. The Eocene, when volcanoes sprouted amid tropical forests. The Oligocene, when the global climate cooled and bombastic calderas held sway. The Miocene, a time ruled by flood basalts and explosive volcanism. The Pliocene, when lakes first inhabited eastern Oregon valleys. The Pleistocene, when glaciers advanced into Washington, and cataclysmic floods washed across the landscape. The Holocene, 11,700 years ago until about AD 1870, before humans became really, really good at pumping the carbon stored in geologic systems for 4 billion years back into the atmosphere. And finally, the Anthropocene, from AD 1870 into the future, a time when humans became a geological force, changing climate, ocean chemistry, and cycles of sedimentation, among many other things.

In this book, you'll notice some significant revisions to our stories of the Northwest. Within the past decade, many understandings have changed. The Steens basalts are now the earliest of the Columbia River basalts (or, more properly, Columbia River Basalt Group) rather than a rogue outlier. A smattering of volcanic buttes in central Oregon are now the ramparts of a supervolcano—the Crooked River caldera. Ice Age Floods, once thought to be restricted to about 15,000 years ago, actually started sweeping across the Palouse more than 700,000 years ago. We have more precise dates for things, including the extinction of dinosaurs (66.038 ± 0.025/0.049 million years

Ice Age Floods carried huge, ice-rafted boulders from Canada and Montana to the Willamette Valley and other destinations. This argilite, enshrined at Erratic State Wayside near McMinnville, Oregon, came from British Columbia.

ago according to Sam Bowring and colleagues at MIT). Importantly, there is increasing understanding that the other four major extinctions (or five, if you count the current, ongoing extinction) were related to rapid climate change. These and many other updates are not the results of previous errors, but of new technologies, more detailed investigation, and occasional serendipitous discoveries.

It was the novelist Marcel Proust who best characterized science, and especially geology, almost a century ago. His statement? "The greatest voyage of discovery is not seeking new landscapes, but having new eyes."

The days of great exploration and discovery of new horizons seem finished. Lewis and Clark are gone. Today, well-worn trails transect wilderness areas. ATV tracks crisscross remote backcountry. It seems there are no truly undiscovered places in the Pacific Northwest.

Or are there?

The timeworn landscape beneath our feet can be just as much undiscovered territory as it was before eighteenth-century fur trappers or the first pre-Clovis mammoth hunters explored the place. As a graduate student in the 1980s, I learned that Smith Rock, a cathedral of tawny cliffs that enshrine the Crooked River 30 miles north of Bend, Oregon, was the product of a fizzy, but demure, volcanic eruption. A separate diminutive volcano crafted Powell Buttes, 25 miles to the south. Ditto for Grizzly Butte, 15 miles northwest. All these,

my knowledgeable professors explained, were separate volcanoes, and nothing extraordinary.

But in 2006, things changed. Called in to explain a mysterious paucity of groundwater for local developments, two geologists from the Oregon Department of Geology and Mineral Industries, Jason McClaughry and Mark Ferns, discovered that all these disparate small vents were actually part of a single, huge volcano—a supervolcano, in fact—that erupted with unimaginable fury 29.5 million years ago.

Smith Rock still looks pretty much like it did in the 1980s. But the meaning of the place has changed completely. Now when I examine the tawny cliffs along the riverbanks, I envision very different things than I did—or anyone did—a decade ago. This newly discovered supervolcano, 25 miles in diameter, Yellowstone in size, and incomprehensible in power, is a newly discovered landscape. Until McClaughry and Ferns unveiled it in 2006, no one knew that the Crooked River caldera, Oregon's biggest volcano, existed. This is just as much a new discovery as if Lewis and Clark themselves had defined this huge volcano.

Discovering the unrecognized landscape already before your eyes ranks as one of the most exquisite thrills science has to offer. Although much of our new knowledge comes to us courtesy of improved technology, much depends on the old-fashioned skill of observation. The landscape, and all its past, awaits us.

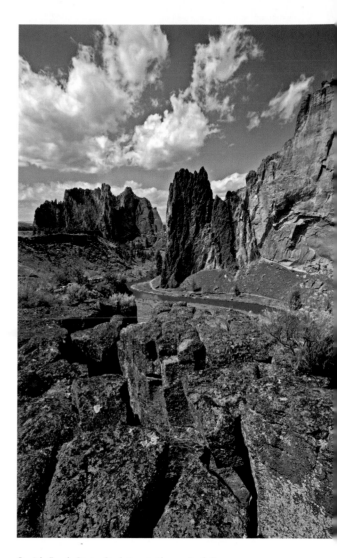

Smith Rock State Park in north-central Oregon was not recognized as the northern ramparts of a Yellowstone-sized, flat-lying caldera volcano until 2006.

The summit of Steptoe Butte, pictured here, is composed of 1.2- to 1.4-billion-year-old quartzite. It rises above the surrounding Washington Palouse and is the type location for a "step toe," or hill of older rock that protrudes above younger topography. The quartzite (foreground) contains zircons that date to about 1.9 and 2.5 billion years in age, based on work by J. Ellis and colleagues.

CHAPTER 2   The First Lands   *Ancient Supercontinents*

Long ago, before there were trilobites or tyran-nosaurs, clams or coelocanths, the earliest foundations of the Pacific Northwest collected along the equator. Life consisted of blue-green algae and some experimental single-celled animals. The land was bare; the sea was green. Oxygen composed less than 1 percent of the atmosphere. Continents covered less than 10 percent of the global surface.

The oldest formations of the Northwest bear witness to these times—an eon known as the Archean, 4 billion to 2.5 billion years ago. The most barren and battered bed-rock of northeastern Washington and the Idaho Panhandle represents the shores of an ancient landmass. Known as Kenorland, after the Kenora mining region in northwestern Ontario, the ancient supercontinent's rocks would become the cores of South Africa, Australia, Scandinavia, and North America. Kenorland was one of many supercontinents that have assembled, broken apart, and then recombined again at a different place, in a different configuration. Kenorland straddled the

Earth's equator. It was a barren land of rocks and dunes and chartreuse seas. Think of Mars with water and maybe some algae.

Two outposts of Kenorland are exposed in the Northwest. They are part of the Priest River complex and Pend Oreille Gneiss, which appear south of Newport, Washington, and in reclusive outcrops near Priest River, Idaho. These stones are banded and deformed. Like the wrinkle-riven faces of the aged, the ancient rocks display multiple signs of stress, weather, and endurance. Research by Ted Doughty of Exxon indicates the Pend Oreille Gneiss was originally a granite-like rock known as tonalite—a rarity in the Archean, when true granite was de rigueur. Related rocks, dark green and laced with salmon-pink feldspars, may have at one time been darker gabbros. Rocks almost as old, and possibly related, make timid appearances beneath the northern skirts of Mount Spokane. These, too, are a foundation of the Priest River complex.

Elsewhere in the Northwest, there are even fainter echoes of the Archean crust. The gran-ites in the northern, 66- to 54-million-year-old

Bitterroot lobe of the Idaho batholith contain zircons. These tiny, durable minerals have a high concentration of uranium and thus are useful for radiometric dating. They yield ages of 2.5 billon years—evidence that the much younger granites were formed, in part, by melting very ancient crust. In the Dishman Hills near Spokane, gneiss that was originally sandstone also yields 2.6- to 2.5-billion-year-old zircons.

Beginning about 2.45 billion years ago, vol-canoes and plate motions tore Kenorland apart. This event marked the end of the Archean eon, and the beginning of the Proterozoic eon, when life strengthened its grip on the planet. Algae, bacteria, and single-celled animals thrived in the sea; the land likely hosted biological crusts—a complex community of blue-green algae, or cyanobacteria. In shallow marine waters, 10-foot-tall mounds of cyanobacte-ria known as stromatolites developed. Their descendants flourish today, in the tidal shallows of Sharks Bay, Australia.

The Priest River Complex contains rocks that are more than 2 billion years old and highly deformed. They likely were part of the crust of Kenorland—the first true supercontinent.

However innocuous their appearance, stromatolites' (and biological crusts') effect on the Proterozoic planet was profound. At 2.6 billion years ago, oxygen made up less than 0.01 percent of the Earth's atmosphere. But by 2.4 billon years ago, riotous photosynthesis by stromatolite/biological crust cyanobacteria and blue-green algae had increased atmospheric oxygen almost hundredfold, to about 1 percent. The result? Iron combined with oxygen and precipitated iron oxides from seawater, transforming the sea from a drab olive-green to a more familiar azure blue. This event, known as the Great Oxygenation Event,

ushered in the Earth's oxygen-dependent life systems. However, it's also termed the Oxygen Catastrophe because any anaerobic life, to which oxygen was toxic, would have vanished. This shift also decreased the relative amount of greenhouse gas, triggering the Huronian Ice Age, which lasted generally from 2.4 to 2.1 billion years ago, with its most intense stage possibly covering most of the globe with ice for more than 60 million years. Great Event or Catastrophe—it's all in your point of view.

The Nuna and Rodinia Supercontinents
Continental landmasses began coagulating again about 2 billion years ago, constructing a new, more massive continent known as Columbia (for the presence of these rocks near the Columbia River's headwaters in British Columbia) or Nuna (for Northern

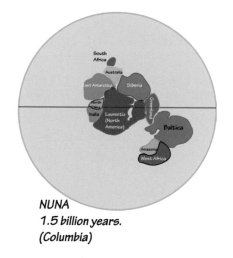

NUNA
1.5 billion years.
(Columbia)

The first supercontinent, Columbia, formed about 2 billion years ago and included the cores of today's continents.

Europe and North America, where the rocks are well exposed) appeared. Here, the Archean granites of Kenorland—now stretched, crumpled, heated, and reconstituted into gneiss—served as the foundation for younger, but still ancient, sediments.

The earliest sediments deposited atop the Archean gneiss appear today as 1.8- to 1.6-billion-year-old metamorphic rocks in Washington's Dishman Hills near Spokane and also on the southern flanks of Mount Spokane. They include the Gold Cup Quartzite—a stretched-pebble conglomerate about 1.4 billion years in age, with dates of included mineral fragments of 2.6 to 1.6 billion years old; the Hauser Lake Gneiss—sediments dated at 1.6 billion years old; and the LeClede Augen Gneiss, "only" 1.5 billion years in age. The summit of Steptoe Butte, north of Pullman, exposes 1.4-billion-year-old sandstone—metamorphosed into its harder cousin, quartzite.

The Belt Supergroup and the Shallow Sea
The rift on the western edge of equatorial Nuna became a broad and relatively shallow, brackish sea that stretched from western Montana north to Alberta and British Columbia, south into Wyoming, and west beyond Sandpoint, Idaho, to Chewelah, Washington. Into this seaway, rivers deposited sands and silts easily pillaged from the still-barren land.

The result was a thick accumulation of fine-grained sedimentary rocks at the bottom of this nameless sea. Today these rocks are known as the Belt Formation—or more properly, the Belt Supergroup. Named for the small town of Belt, Montana, the Belt Supergroup includes limestones, sandstones, and shales. Belt rocks reach their maximum thickness of 53,000 feet (10 miles) in north-western Idaho.

The sedimentary rocks of the Belt Group are virtually unmetamorphosed. In Montana, where shallow waters predominated, limestones rule. In western Idaho where water was deeper and life more precarious, shales and sandstones predominate. Near Chewelah, Washington, the westernmost exposure of this ancient shallow sea, Belt Supergroup sediments (part of the "Deer Trail group" and Buffalo Hump Formation) are an estimated 20,000 feet thick, and consist mostly of siltstones burnished by the faintest of metamorphic heat.

The sedimentary rocks of the Belt Supergroup reveal much about the Northwest's environment 1.2 to 1.5 billion years ago. In some, impressions of raindrops remember passing storms. In others, mud-cracks captured the heat of an ancient sun. Ripple marks record the directions of long-vanished currents. Along Going to the Sun Highway, mounds of stromatolites remember the placid shallow waters of a brackish bay. If you can imagine Chewelah, Washington, as looking like the desolate landscape along the Red Sea you have a pretty good vision of what the Nuna landscape might have looked like.

When eruptions and newly invigorated plate motions slowly tore Nuna apart about 1.4 billion years ago, the Belt Basin persisted,

keeping much of North America joined tenuously to Australia, while the cores of Africa, South America, Siberia, and Europe drifted away. These errant fragments of Nuna remained dispersed for about 300 million years. Then, the continents collected once again into another—and more famous—supercontinent about 1.1 billion years ago.

Rodinia: The New Supercontinent

The new supercontinent, amalgamated about 1.1 billion years ago, is called Rodinia. The word Rodinia—from the Russian infinitive *rodit*, meaning "to beget" or "to grow"—was chosen in the 1970s when it was thought that Rodinia gave birth to all subsequent continents and its edges served as loci for the development of complex animals. Rodinia lay mostly in the southern hemisphere, though reconstructions vary. It contained the cores of today's continents. Paleomagnetic reconstructions suggest that eastern Washington and northern Idaho likely lay near the South Pole. Antarctica was at the equator. What would become Idaho and western Washington was barren real estate that probably looked a lot like the Arizona desert, minus the cacti.

Although plucky life-forms plied its shore-lines, and likely pioneered the land, Rodinia was a primitive, hostile world. Atmospheric oxygen was less than 10 percent of today's levels, and there was not enough ozone to block ultraviolet radiation. Much of the landscape was flat. There was little vegetation to break the force of rain. No roots to hold soil. Like Kenorland and Nuna before it, the landscape

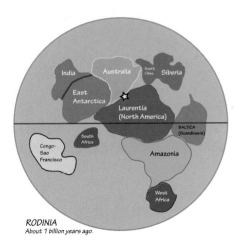

RODINIA
About 1 billion years ago.

Rodinia (from the Russian "Родина," *ródina,* meaning "The Motherland") existed between about 1 billion and 750 million years ago.

of Rodinia was a bit like today's Mars, but with water and oceans, prolific algae, and biological crusts. The only rocks in the Northwest that were deposited in Rodinia's shallow waters are found in the Buffalo Hump Formation—exposed on Buffalo Hump north of Chewelah, Washington, and composed, ostensibly, of fragments of the Belt Group sandstones.

Ultimately, Rodinia would rift apart about 700 million years ago. In what would become the Northwest, the continent broke along a line that leads through Sandpoint and Priest River, Idaho, creating the first Northwest coastline and scattering the other half of the Priest River complex around the globe. Today these missing rocks appear in the Georgina Basin, Grawler pluton, and Jervois Range of central Australia, and in the Udzha Basin of eastern Siberia.

The Selkirk Mountains of northeast Washington sequester greenstones and sediments deposited during Snowball Earth.

CHAPTER 3   A Planet In Crisis   *Snowball Earth*

Were it not for Rodinia's eventual breakup into early versions of today's landmasses, life on our blue planet might have expired more than 700 million years ago.

Strong evidence indicates that the Earth was encased in a frigid Ice Age during much of the late Proterozoic time, about 780 to 630 million years ago. It was a time hostile to life, when the fossil record dwindles to next to nothing. The time period is officially dubbed the Cryogenian. But most geologists just call it "Snowball Earth."

Snowball Earth's trigger remains a mystery. Hypotheses include the polar distribution of continents, massive eruptions of ash that shielded the planet from sunlight, or the sequestration of greenhouse gases by rapid geochemical weathering. Calculations and computer models by William Sellers of the University of Arizona and Mikhail Budyko of Russia's Main Geophysical Observatory, demonstrate that once glacial ice covers about 35 percent of the globe, the increased reflectance of sunlight and heat (albedo) becomes a feedback loop that forces additional cooling, and the planet ultimately becomes a glistening, white, ice-covered freezer.

During the 150 million years of Snowball Earth, the planet cycled between extreme cold and extreme heat, icehouse to greenhouse, at least twice and probably three times. Major glacial episodes, in which all or most of the planet was encased in ice, peaked 716 million years ago, a time period known as the Sturtian glacial epoch, when glaciers extended to the equator. The Marinoan glacial epoch, a second though less complete global glaciation, occurred 640 million years ago when glacial ice covered all land areas and most of the northern and southern hemispheres. An early cold period, from about 780 to 740 million years ago (the Kiagas), may have produced global glaciation as well. During the interspersed times of extreme warmth, temperatures likely topped 120 degrees F, and CO_2 levels exceeded 350 times today's concentrations.

Ultimately, volcanism rifted Rodinia apart, spewing greenhouse gasses into the atmosphere and rescuing the Earth from an icy tomb. Outcrops of greenstones on Round Top Mountain, Green Mountain, and Thunder Mountain northeast of Metaline Falls are vestiges of the basalts that eventually splintered the supercontinent. Karen Lund has dated them at about 685 million years in age. Atmospheric CO_2 unleashed by these eruptions boosted global temperatures. As continents dispersed, the conditions that allowed cycling between climatic extremes vanished.

In the Northwest, several formations help tell the tale of Snowball Earth. Idaho bedrock, including the Pocatello Formation in southeastern Idaho and the Edwardsburg Formation in central Idaho, include glacial tills with ice-grooved rock clasts. These rocks span the time of 716 to 667 million years ago,

Proterozoic greenstone on Roundtop
Mountain east of Metaline Falls,
Washington, is a remnant of the basalts
that both tore Rodinia apart and rescued
the planet from its icy encasement.

Volcaniclastic breccias at the base of Round Mountain
include volcanic tuffs and angular fragments. These rocks
hint at the unrest that eventually ended Snowball Earth.

the Sturtian glacial epoch. Rocks exposed on Scott Mountain, near Pocatello, Idaho, include a conglomerate of smoothed and striated stones. The grooves on the clasts could only have been gouged by glacial ice. In northeast Washington, glacial footprints are etched into the Shedroof Conglomerate, Huckleberry Conglomerate, and Leola volcanics.

The Northwest harbors evidence of the warm periods as well as severely cold ones. In central Idaho, the Moores Station Formation, near the top of Marshal Mountain, is mostly limestone—or now, its metamorphosed equivalent, marble. These limestones are the products of a suddenly warmer world, a place of high temperatures and abundant dissolved carbon that precipitated out again as calcium carbonate—a chemical limestone.

End of the Snowball: Life 2.0

During Snowball Earth, life faced a grave crisis. Planetary life consisted of diaphanous algae, stromatolites (also algae, or more properly, cyanobacteria), and possibly some experimental eukaryote cells (cells with mitochondria, chromosomes, and other trappings of multicellular organisms, including animals). Life needed light, it relished warmth, and in a frigid, ice-covered world, sunlight was hard to come by.

Marine life may have been nurtured by a superabundance of nutrients, according to Noah Planavsky and colleagues of the University of California, Riverside. They found that the Snowball Earth seas were enriched in phosphorus—a result of huge amounts of sediments that glaciers ground into fine clays and carried into the sea. In a sort of Rube Goldberg effect, high phosphorous nourished algae, which produced more oxygen, and took up carbon dioxide, both accelerating the planetary cooling, and, for the first time,

providing a relatively oxygen-rich atmosphere (perhaps 10 percent versus today's 22 percent) that would support animals. The stress—and the opportunities—provided by a changing atmosphere led to multicellular functions in groups of eukaryote cells, and ultimately, the rise of metazoans—multicellular animals—in refugia somewhere beneath the shroud of ice. Recent discoveries of cold-water-adapted communities beneath the Antarctic's Ross Shelf support the idea that life adapted to the cold and developed more robust forms.

The rifting of Rodinia opened the ancestral Pacific Ocean and established oceanfront property on the western margin of North America. Volcanoes, accretion, and sedimentation would modify this nascent Pacific Northwest margin greatly through time, building the coast farther west. But 700 million years ago, the Pacific Northwest became a place of true maritime effect, a lithic and tectonic distinction that has endured.

The Addy Quartzite, about 540 million years old, represents Washington's first beach, near the town of Addy, in easternmost Washington.

CHAPTER 4   Native Shores   *The Northwest's Foundation*

Rodinia's demise helped banish Snowball Earth forever, and opened the Panthalassa Sea—the ancestral Pacific. Australia retreated westward. Siberia inched north. Sediments collected along the balmy, equatorial beaches of western North America (Laurentia). This new shoreline extended from today's Death Valley and western Arizona north through central Idaho and eastern Washington and along the modern uplifted spine of British Columbia. The nascent ocean became a nursery—first for tiny sowbug-like trilobites, small globular sponges, and a variety of marine worms; and later, for even more bizarre-but-robust animals—the Burgess Shale fauna. The early coast of eastern Washington, western British Columbia, and western Idaho, as well as islands offshore, were valuable cradles for life's progression.

The Earliest Beach

In Washington's northeast corner, and in Idaho and British Columbia's Selkirk Mountains, the ancient shore persists. It includes quartz-rich sandstones deposited during rifting, and limestones that bear the abundant fossils of primitive trilobites and tall, conical sponge-like animals called archaeocyathans. These pristine beaches represent a new order, and the development of complex life.

The oldest and most mysterious lives are those of the Ediacarans—rare, diaphanous forms that appeared about 630 million years ago and vanished by 540 million years ago. Although they seem to represent an evolutionary dead end, they were apparently the first major foray into multicellular, multifunctional organisms. However, the Ediacaran's exact nature, and even kingdom, is controversial. Fossils of the 1/2-inch-long, horseshoe-shaped *Kimberella* suggest a creature with bilateral symmetry similar to modern mollusks. But Greg Retallack, of the University of Oregon, considers many Ediacarans to have been mosses or algal colonies that lived in ancient soils. Other paleontologists suggest they were neither animal nor plant, assigning them their own unique kingdom—the Vendobionta. Whatever their nature, the Ediacarans persisted for almost 90 million years—as long as the time from tyrannosaurs until the time you are reading this—not a bad record for creatures the size and dimensions of (but far more fragile than) the loose change in your pocket.

In the Northwest, Ediacaran fossils are found only in the Yreka Formation in the California Klamath Mountains where basaltic seafloor of similar age is preserved in the 550- to 579-million-year-old bedrock of the Trinity Alps. In the Yreka Formation, geologist Nancy Lindsey-Griffith has identified the diaphanous fossils as *Beltanella* and another species of *Ediacaria*. Their age is broadly pegged at 640 to 575 million years.

The Yreka Sandstone is part of the Eastern Klamath terranes—offshore continental fragments, seafloor, or islands added to North America when the westward-moving continent collided with them. These sandstones contain diaphanous Ediacaran fossils—small, fragile animals (?) that lived about 600 million years ago. Lichen in bottom left corner for scale.

The equivalent rocks of northeastern Washington (Addy and Gypsy Quartzites) may be too metamorphosed to have preserved the delicate forms of Ediacarans. Still, they may have been here. At the time Ediacarans flourished, the warm beaches of northeast Washington were clean, glistening quartz sand—now preserved in the Gypsy Quartzite. These rocks, 560 to 540 million years in age, straddle the Cambrian–Precambrian boundary. They form the barren slopes of 7,320-foot Gypsy Mountain in the Selkirks.

The uppermost part of the Gypsy Mountain Quartzite unit includes limestones—the sediment of warm and supremely quiet waters. Trace fossils include elongate "burrows" that may represent a burrowing, worm-like creature—a rare fossilized inhabitant (*Scolithus linearis*) found just north of Sullivan Mountain.

The Addy Quartzite—exposed in roadcuts just west of Addy, Washington—represents a slightly younger but still pristine beach, 540 to 530 million years old. These rocks are blatantly Cambrian. In outcrop, the individual beds of sandstones (now metamorphosed into quartzite) retain their bedding. From a distance, they look like the original sandstones. Some bear the faint impressions of trilobites. Only two trilobite varieties are known from the Addy Quartzite—*Nevadia addyensis* and *Esmeraldina argenta*. They are tiny. The shield, or head of *N. addyensis* measures just 1/4 inch wide. Brachiopods (clam-like mollusks) also appear in the Addy Quartzite. Although their identity is indeterminate because of metamorphism and deformation, their presence, along with trilobites, demonstrates that life flourished on Washington's very early Cambrian shore.

The Cambrian Explosion and the Early Paleozoic: Life Proliferates

As you drive along Washington Route 31 into Metaline Falls, you will find a plethora of layered gray rocks exposed in roadcuts. Some yawn with mine shafts excavated in the early 1900s; others host quarries. About 500 million years ago, these slightly metamorphosed limestones were a warm seafloor, the lair of trilobites and reefs built by sponges. Today these rocks are the economic lifelines of the community. Welcome to the Metaline Limestone.

At Metaline Falls, limestones expose a marine ecosystem of the Middle Cambrian age—about 520 million years old. There were algae and sponges galore in these seas. And at least twenty kinds of trilobites scuttled across the limey sea bottom. Like their predecessors in the Addy Quartzite, they were not very big. *Elrathia longiceps* measured about a 1/2 inch long; *Olenoides maladensis* grew to almost 1 inch in diameter.

Younger shales of the Maitlen Formation overlie the Metaline Limestone. These shales are the contemporaries, and perhaps even an extension of, British Columbia's renowned Burgess Shale deposit 300 miles to the north. Fossils similar to the iconic Burgess fauna have not (yet) been found in Washington or Idaho.

The charismatic fossils of the Burgess Shale perch on mountain cliffs at Yoho National Park in easternmost British Columbia. Here, unmetamorphosed sedimentary rocks, 515 million years in age, contain fossils of bizarre soft-bodied animals, including a weird, spikey sea cucumber–like organism aptly called *Halucogenia*, and the first true marine predator, the 3-foot-long *Anomalocaris*—a giant of the time. With two long "arms" for grasping, and a round mouth of fearsome toothy plates, *Anomalocaris* was more oddly horrific than anything yet devised by Hollywood computer graphics.

The Burgess Shale fauna developed in shallow ocean waters sheltered by a limestone reef. Their soft bodies are preserved here due to rapid burial in oxygen-poor conditions, with seawater that was high in calcium carbonate and low in sulfur, in sediments that mercifully escaped metamorphism, according to Robert Gaines of Pomona College. Fossils of similar but slightly older (527 million years) Burgess Shale creatures have been discovered in China and Greenland. Similar animals may have lived in equivalent waters of Washington's Maitlen Shale, but been erased by time and temperature.

The reason for the sudden Cambrian explosion of complex life remains a mystery. Perhaps it was warming and re-oxygenation of seas after the long Snowball winter. Perhaps seawater composition changed, driven by increased erosion on land. Perhaps somewhere beneath the sea surface, life evolved and blossomed in places now purged from the fossil record, or not yet found.

Limestones—composed of calcium— have three properties useful to mankind.

The Cambrian age Metaline Limestone contains fossils of trilobites as well as early crinoids and brachiopods. Although not strongly metamorphosed, the rocks are deformed.

Maitlen Formation shales, about 505 million years in age, have been metamorphosed into a low-grade metamorphic rock called phyllite in exposures near Metaline Falls.

Gardner Cave, in Crawford State Park, is the second longest cavern in Washington. Dripstones and flowstones decorate the walls. The use of lights has fostered growth of algae, which threaten some cave formations.

First, limestones host caverns. Northeast Washington is no exception. North of Metaline Falls, and literally within spitting distance of the Canadian border, a tiny state park, run collaboratively with the US Forest Service, hosts Gardener Cave—a mile-long cavern discovered in 1892 and now visitor-friendly with lights, stairways, and handrails. Second, the carbonate readily neutralizes the acid fluids that dissolve and transport metals. Limestones are exceptional places for ore deposits. At Metaline Falls, gold was discovered in 1850, but major mines opened about 1910—not for gold, but to extract zinc and lead from the Metaline Limestone. Mining continues today. Third, limestone is an essential ingredient in cement. The Lehigh Portland Cement Company has operated there since 1909.

In Washington's northeastern corner, the hills that flank the Columbia River between China Bend and Northport sequester metamorphosed sedimentary rocks from a slightly younger time. They are known as the Covada Group. Deposited between 485 and 443 million years ago, they represent the Ordovician period, when atmospheric oxygen had risen to about 14 percent, atmospheric carbon dioxide averaged about 4,200 parts per million (ppm), and global seas were toasty. Like the Metaline Limestone, these rocks were deposited on the continental shelf off the Idaho and Montana coastline. Animals were more sophisticated. Many tribes of trilobites prowled coastal waters. Jawless armored fish appeared, along with more complex coral reefs, squid-like cephalopods, and other ecosystem upgrades. The fine-grained, dark rocks among these hills (the Ledbetter Shale) contain fossils of graptolites—a 2-inch-long, free-floating animal whose fossil looks like a saw-toothed blade of grass.

It seemed a placid life on Washington's Early Paleozoic shelf. But that was not to last. The proto-Atlantic Ocean (known as the Iapetus Ocean) began to close, colliding Europe, Scotland, and Scandinavia with North America. The beginning stages of Pangaea were under way. Stresses from the collisions of eastern North America with islands and small continents began first to uplift and fold the quiet shelf and the continental edge, and ultimately, spawn subduction and volcanic activity along today's tectonically placid eastern edge.

Pangaea

In Earth's history, as in our own, changes on one side of the globe affect things far, far away. This seems to be exactly what happened about 480 to 440 million years ago. By the end of the Ordovician time period, 440 million years ago, the ancestral Atlantic (the Iapetus Ocean) had closed. Africa and South America merged into Gondwana—the southern composite continent that was centered on the South Pole. A separate, composite landmass, Laurentia, lay along the equator. Here, Scotland and Greenland collided with what is today the East Coast of the United States (the Taconic Orogeny), while Scandinavia merged with the rest of Europe.

The largest supercontinent, Pangaea, developed when Gondwana and Laurentia merged, about 300 million years ago. A collage of the cores of modern continents, Pangaea extended from pole to pole, changing oceanic and atmospheric circulation and contributing to the globe's greatest mass extinction before its breakup about 200 million years ago.

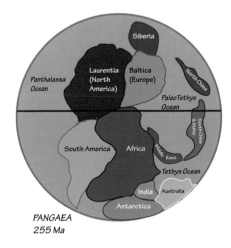

PANGAEA
255 Ma

Pangaea, the last major supercontinent, began to break up about 140 million years ago, sending North America westward, and beginning collisions with islands and sea-floor in the Pacific

The Kootenai "Arc"

The stress of the long-term collisions that formed Laurentia and, ultimately, Pangaea, rippled across Laurentia, ultimately deforming the West Coast. In the Northwest, some formations on the thin, leading edge of the continent were folded and uplifted. They are termed the Kootenai Arc (also variously spelled Kootenay). Seen on a map, the Precambrian and Paleozoic rocks of the Kootenai Arc form a wide curve from British Columbia south toward Chewelah, Washington, where they disappear beneath the Columbia River basalts. The Kootenai Arc is an arc in the geometric sense of the word—a structural warp rather than a volcanic event. Eruptions were minimal. Just across the Canadian border from Washington, at the north end of the Selkirk Range, Ordovician pillow lavas (basalt that congeals into rounded forms when it cools quickly in or under water), tuffs, and volcanic greenstones indicate sporadic volcanic activity, probably in a sort of rift zone.

The North Cascades are composed of multiple exotic terranes added as North America collided with islands, seamounts, ocean ridges, and continental fragments about 100 million years ago.

Global Temperatures

Millions of years — PROTEROZOIC | PALEOZOIC | MESOZOIC | CENOZOIC

Camb Ordo Silur Dev Miss. Penn Perm Trias Juras Creta Paleo Eoce Olig Mioc Plioc Pleist Holo Anth

CHAPTER 5   Terranes and New Territory   *Assembling the Northwest*

The Northwest's history during almost 300 million years of the Paleozoic and Mesozoic is told not by native stones, but by geologic immigrants known as exotic terranes—rocks formed elsewhere and added to North America by plate tectonics. The rocks of these terranes reveal global climates that swung from ice ages to greenhouse tropics, and global life that endured cataclysmic extinctions.

None of the three major extinctions during this time are directly represented in the Pacific Northwest's exotic terranes. But their consequences echo through our geologic record. During the Ordovician extinction 445 million years ago, rapid climate change and shifts in ocean chemistry and temperature eradicated about 60 percent of marine fauna. The cataclysmic end-Permian extinction—now precisely dated at 251.941 to 251.880 million years ago by Seth Burgess and Sam Bowring at MIT—wiped out 96 percent of Earth's biota in a span of just 60,000 years. The cause

seems linked to vast volcanic eruptions in Siberia (the Siberian Traps) that released huge quantities of carbon dioxide, sulfur, and other toxic substances into the atmosphere. In concert with eruptions, microbes pumped prodigious amounts of methane into the atmosphere. A study published in 2014 by Daniel Rothman and others at MIT found that methane-producing microorganisms called *Methanosarcina* made a sudden, genetic change 252 million years ago, allowing it to transform CO_2 dissolved in sea water into atmospheric methane—a potent greenhouse gas. The double whammy of volcanic and biologic greenhouse gas and other toxins produced catastrophic shifts in climate and ocean chemistry and very nearly ended life on the planet. Recovery required more than 10 million years, and gave us the foundational genera of both dinosaurs and mammals. Just as things seemed on an even keel, at 201.3 million years ago the Atlantic Ocean began to open, producing voluminous eruptions and triggering the third-largest extinction

in planetary history, eradicating half of all known marine species, weakening many reptilian genera, and paving the way for the rapid rise of dinosaurs.

The opening of the North Atlantic Ocean began the breakup of Pangaea. North America began to move westward. By the middle of the Jurassic, 180 million years ago, volcanic islands and continental snippets, ocean crust and marine plateaus, all began the lengthy process of collision with the coasts of Idaho and Utah. These exotic terranes would not become fully part of the continent until well into the Cretaceous, 120 million years ago. Today, they are the foundation for most of Oregon, Washington, Northern California, and virtually the entire west coast Cordillera, from Baja California to Alaska.

Indonesia is the modern equivalent of this ancient, island-dotted Panthalassa Sea. Today, the seas between Australia and Asia harbor hundreds of islands and multiple subduction zones. Volcanoes jut from the sea to heights of 10,000 feet and more, spewing ash, debris flows, and lavas with explosive violence. (Think Krakatoa.) Plates are sliced apart by faulting. Geology is chaotic. Rocks on the Philippine islands, notes Michael Hamberger and his colleagues at Cornell University, are "a jumble of arc magmatic units, subduction mélange, sediments shed from volcanoes, fragments of continental crust rifted from Asia, and depositional basins of various ages."

This is also a fitting description of geology in the North Cascades, Wallowas, San Juans, Klamaths, and northern Sierra Nevada. All are composed of "exotic terranes"—rocks that were formed as islands and ocean floor and were plastered onto the western coastline as North America began its westward trek.

Exotic Terranes: The Concept

Exotic terranes are geologic immigrants. Born elsewhere, they travel hundreds or thousands of miles on the sturdy backs of tectonic plates. They arrive on new shores as discrete individuals, with a history and character all their own. Some may be mature, with long resumes and diverse experience. Others may be young, energetic, and lean. Some arrive as a whole family, with aged grandparents and newborns, others as individuals. And like human immigrants, all contribute to the diversity, prosperity, and function of the new land.

The spelling "terrane" is used to distinguish a geologic landscape, or set of rocks, from the more conventional geographic term "terrain." Geologic terranes preserve a stratigraphy that

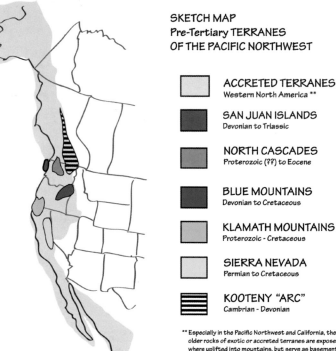

SKETCH MAP
Pre-Tertiary TERRANES OF THE PACIFIC NORTHWEST

- **ACCRETED TERRANES**
  Western North America **
- **SAN JUAN ISLANDS**
  Devonian to Triassic
- **NORTH CASCADES**
  Proterozoic (??) to Eocene
- **BLUE MOUNTAINS**
  Devonian to Cretaceous
- **KLAMATH MOUNTAINS**
  Proterozoic - Cretaceous
- **SIERRA NEVADA**
  Permian to Cretaceous
- **KOOTENY "ARC"**
  Cambrian - Devonian

** Especially in the Pacific Northwest and California, the older rocks of exotic or accreted terranes are exposed where uplifted into mountains, but serve as basement covered by younger rocks over the remaining area.

Map showing exotic terranes in the Pacific Northwest. From Baja California to Alaska, the foundation of the west coast of North America is terranes added to the continent in the Cretaceous or later. The exotic terranes of the PNW (and elsewhere) are exposed where they have been uplifted.

FACING PAGE: The San Juan Islands, viewed from Mount Moran, are a western extension of the terranes exposed in the North Cascades. They remind us of their origins mostly as islands in the Pacific (Panthalassa Sea) off the Idaho shore.

may be sedimentary, metamorphic, and/or igneous. Each has a geologic history distinct from surrounding terranes or the continental margin. How can we tell that these rocks originated elsewhere? The evidence comes from three sources.

First, there are the rocks themselves. Many terranes in the Northwest contain rocks that could only have been deposited in tropical settings. Limestones, with their corals and warm-water sponges, are scions of warmer places than the Wallowa Mountains or the North Cascades.

Second, fossils in many exotic terranes match fossils found far, far away from the Northwest. The Triassic ichthyosaur found in the Wallowa Mountains, for example, has kin in both northern California and China, as do the corals and brachiopods associated with the ichthyosaur. The presence of pedestrian mollusks and corals in both China and Australia suggests that the rocks were originally deposited at locations much closer together, and have been moved by the magic of plate tectonics.

The clinching argument for the exotic nature of terranes comes from an exotic science—paleomagnetics. When sedimentary rocks are deposited, or lavas cool and solidify, tiny crystals of iron oxide (magnetite) align themselves with the Earth's magnetic field. Their orientation reveals the latitude—how far north or south of the equator—in which they formed. If the rocks then move north or south, the iron oxides still retain that original, telltale signature of their birthplace,

which geophysicists can detect and measure. Paleomagnetic data provides information about how far from the equator rocks were formed, although we cannot yet resolve whether they were north or whether they were at the same latitude south of the equator based on this measurement alone. Still, paleomagnetic results have confirmed and refined earlier estimates based on lithologies and fossils.

Collisions Build the Northwest's First Real Estate

The exotic terranes that originated west of Idaho were never far from the North American continent, according to Todd LaMaskin of the University of Oregon. Tiny minerals known as zircons that are present in sandstones of the exotic terranes are more than a billion years old. They could not have come from the much younger volcanoes or seafloor of the exotic terranes themselves. Instead, they were eroded and transported from North America (Idaho, Wyoming, or Utah) where older rocks are abundant. Thus, most of the exotic terranes formed developed close to North America, rather than far offshore.

The Klamath Mountains are the southernmost of the accreted (exotic) terranes in the Northwest. Their geology is complex. The multiple terranes are broadly divided into groups, or "belts": the Eastern Klamath belt (Early Paleozoic to Jurassic), Central Metamorphic belt (Devonian), Western Jurassic belt, Western Paleozoic and Triassic belt, and Franciscan belt, mostly Mesozoic in age. Each of these is composed of multiple "terranes," with their own distinctive origin. So, for example, the Cretaceous Franciscan belt, named for its kindred rocks near San

Francisco, includes the Gold Beach terrane (sedimentary mélange), Picket Peak terrane (sediments and serpentinite-matrix mélange), and the Sixes River terrane (mudstones). These individual terranes joined together into the Franciscan belt prior to accretion to North America.

The Western Paleozoic and Triassic belt is subdivided into three broadly related terranes: Rattlesnake Creek (upper mantle rocks that include some chrome deposits), Hayfork (andesites of an island arc), and North Fork (oceanic and upper-mantle rocks). Together, these terranes represent parts of an island arc that collided with North America.

Perhaps the most intimate view of the Western Paleozoic and Triassic belt can be found at Oregon Caves National Monument. Oregon Caves is a small but intricate cavern

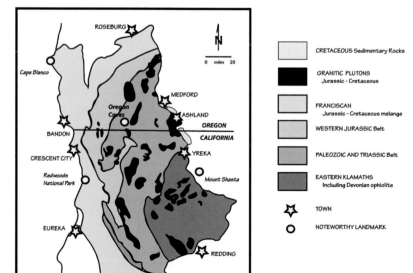

The Klamath Mountains of southwest Oregon and northwest California harbor a complex, intricate assemblage of exotic terranes.

system dissolved from Triassic marble. Once, 230 million years ago, these rocks were coral reefs and limey carbonate banks. Post-accretion, about 140 million years ago, deformation, and heating by the intrusion of the nearby granitic Grayback pluton, recrystallized the limestones into marbles. The openings in solid marble developed as groundwater reacted with organic matter to produce carbonic acid. The weak organic acid slowly dissolved the marble, producing the cavern.

Thomas Condon, Oregon's pioneering geologist, unintentionally demonstrated the slow but steady accumulation of calcium carbonate within the cave. In 1883, Condon and a group of students visited and explored the cave. At one point, about 700 feet into the cavern, Condon took a pencil and wrote his name and the names of his students on the cave wall. (What prompted this graffiti by an eminent scientist is unknown.) Today those signatures are covered in a protective coat of (relatively) clear calcite—so well protected that when the National Park Service decided to return the cave wall to its natural condition, the pencil marks couldn't be erased without damaging the surface. So Condon's graffiti remains, a signature still visible to visitors, and one of his many lasting legacies.

The Western Jurassic belt hugs southeast Oregon's coastline. It includes the Josephine ophiolite (oceanic crust), exposed at Chetco and Vulcan Lakes east of Brookings, Oregon, and the Rogue-Chetco volcanic arc complex exposed along much of the Rogue River.

Flowstones and draperies decorate the ceiling of the Paradise Lost Room in Oregon Caves.

Thomas Condon, Oregon's pioneering geologist, left his signature and the names of students on his expedition in the cave in 1883. They are still visible beneath a protective coating of clear calcite.

Together, the rocks of the Western Jurassic belt represent a Late Jurassic ocean basin (160 to 165 Ma) that developed as a rift in the older arc, before it was accreted to North America.

In northeast Oregon, another series of terranes are known broadly as the Blue Mountain Island Arc. They include the Wallowa terrane (island arc), the Baker terrane (subduction zone), and the Olds Ferry terrane (a younger island arc). Compared with the Klamaths, the Blue Mountains present a refreshingly homogeneous geology that is perhaps the equivalent of a single Klamath belt.

The Wallowa terrane represents a long-lived island arc system that includes two distinct sets of volcanic islands—one Permian, 270 to 260 million years old, another Triassic, about 235 to 225 million years old. The arc rocks and associated sediments are best exposed in the Wallowa Mountains and Hells Canyon. The oldest rocks in Hells Canyon—280-million-year-old gneiss of the Cougar Creek Complex—are found just south of Pittsburgh Landing, and are well exposed at historic Kirkwood Ranch. In Hells Canyon, the basalts and andesites erupted from arc volcanoes pose as metamorphic greenstones. The seafloor around the island arc forms the limestones of the Martins Bridge Formation and fine-grained sandstones of the Lower

Sedimentary Series and Hurwal Formation, all found in the Wallowa Mountains. These rocks compose the colorful slopes of Eagle Creek, Chief Joseph Mountain, and other peaks. Sacagawea is a mountain of marble—Triassic limestone metamorphosed by the intrusion of the much younger "granites" of the Wallowa batholith. The Matterhorn's precipitous, glacially carved west side exposes marble and metamorphosed seafloor shales.

A younger Triassic to Jurassic set of volcanic rocks and sediments, the Olds Ferry terrane, lurks to the south near the town of Huntington. These volcanic rocks have a significantly different composition compared with the Wallowa terrane, and suggest that the Olds Ferry terrane was an island arc with more explosive, silica-rich volcanoes.

At Pittsburgh Landing in Hells Canyon, the Jurassic Coon Hollow Formation—relatively unmetamorphosed Early to Middle Jurassic shales, limestones, sandstones, and conglomerates—document a progression from freshwater sediments to marine, as Jurassic sea levels changed. Conglomerates and crossbedded sandstones are alluvial or river deposits, produced in a braided stream or meandering river environment. Marine fossils include ammonites and corals, which date these rocks as Middle Jurassic—about 176 to 161 million years ago. The fossil leaves and stems of land plants, including ferns, lycopods, horsetails, conifers, and ginkgoes, also occur in the Coon Hollow Formation.

The Wallowa Mountains of northeastern Oregon expose the Triassic and early Jurassic sediments that were part of an island arc system.

The orange-brown-colored rocks near Chetco and Vulcan Lakes in southwestern Oregon are peridotite—remnants of the upper mantle and Jurassic seafloor now uplifted into Klamath peaks.

Where there are island arcs, subduction zones convey the seafloor deep into the mantle—ultimately generating island arc volcanoes far above. In the Blue Mountains of northeast Oregon, two remnants of subduction zones are known as the Baker and Grindstone terranes. In these chaotic formations, there is no normal stratigraphic order. They are known as mélanges—French for "mixture." In some places, sediments envelope older sediments and seafloor. In others, especially the Greenhorn Mountains, polished, shiny green rocks known as serpentinite envelop chunks of seafloor rocks that were dragged into the maw of subduction. Serpentinite is an altered mantle rock (peridotite) that is less dense than the surrounding mantle and shoulders its way toward the surface, picking up a great variety of blocks from the surrounding crust and subduction zone on its way up. These remnants of reefs, pillow basalts, and fragments of oceanic islands poke out of the serpentinite matrix like large raisins in a pudding.

Another outlier of the Baker terrane mélange appears north of Mitchell, Oregon. Here, a very small area of mélange with blueschist—a fine-grained metamorphic rock produced only in the high pressures and low temperatures of subduction zones—and marble is exposed at Meyers Canyon.

A similar though much more complex story is evident in the North Cascades. There, the accreted terranes are divided into three "domains," similar in concept to the Klamath's "belts." Each domain has a distinctive style of geology. The three domains have been juxtaposed by faulting along the (now inactive) Strait Creek and Ross Lake Faults.

The Western Domain includes terranes of subduction and island arc origin (mélange) like the Belle Pass terrane. There, a block of Precambrian gneiss (The Yellow Aster Formation) equally huge fragments of mantle and younger fine-grained oceanic sediments.

The Blue Mountains' exotic terranes likely consist of several island arcs and subduction zones.

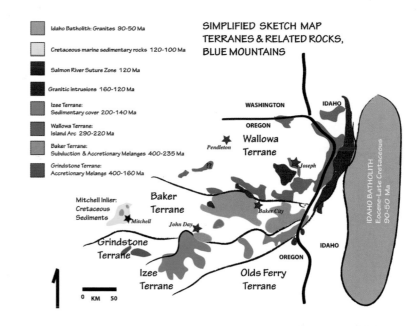

SIMPLIFIED SKETCH MAP
TERRANES & RELATED ROCKS,
BLUE MOUNTAINS

Idaho Batholith: Granites 90-50 Ma

Cretaceous marine sedimentary rocks 120-100 Ma

Salmon River Suture Zone 120 Ma

Granitic intrusions 160-120 Ma

Izee Terrane:
Sedimentary cover 200-140 Ma

Wallowa Terrane:
Island Arc 290-220 Ma

Baker Terrane:
Subduction & Accretionary Melanges 400-235 Ma

Grindstone Terrane:
Accretionary Melange 400-160 Ma

In the North Cascades, Mount Shuksan and the Yellow Aster Meadows area are part of the Western Domain—a complex assemblage of many diverse terranes. Yellow Aster Meadows sequesters some of the oldest rocks in the North Cascades.

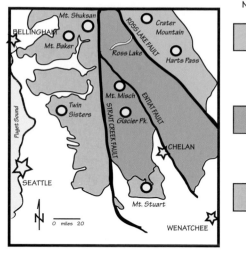

GEOLOGIC SKETCH MAP
NORTH CASCADES DOMAINS

METHOW DOMAIN:
Consists of the Hozomeen Terrane,
Carboniferous to Triassic, 350 to 220
million-year-old sedimentary rocks
and slightly metamorophosed basalts.
Younger, Jurassic to Cretaceous
sandstones and shales, 200 to 95
million years, overlie these older rocks.
Very low-grade to no metamorphism.

METAMORPHIC CORE DOMAIN:
Terranes that are highly metamorophosed
and generally composed of gneiss and
migmatites. Includes the Swakene and
Chelan Mountains terranes, Skagit Gneiss,
Nanson Terrane, and Chiwaukum Schist.
Granitic rocks are derived from melted
metamorphic rocks of the domain.

WESTERN DOMAIN:
A folded stack of terranes, including the
Belle Pass melange, the highly
metamorphosed Easton Terrane
(including the Shuksan greenschist and
blueschist), and mildly metamorphosed
Devonian to Triassic Chilliwack River
Terrane.

The complex terranes of the North Cascades are divided into three "domains" with distinctive histories and deformation styles.

Other terranes in the Western Domain represent sediments shed from an island arc (Nooksak terrane), the metamorphosed volcanic rocks of a different arc (Chilliwack River terrane), and remnants of the seafloor (Easton terrane), which includes Mount Shuksan. The original age of Shuksan's greenstones is uncertain, but metamorphism likely occurred in the Permian, about 260 million years ago.

The Metamorphic Core Domain comprises some of the North Cascades' highest and most rugged peaks, including Mount Buckner and Mount Forbidden. Most of these rocks were metamorphosed at relatively high temperatures and pressures. Terranes include the Chelan Mountains terrane—once ocean-floor basalt, now a dark amphibolite, and fine-grained sediment, now a more glittery mica schist. The Little Jack terrane surrounds much of Ross Lake. Once parts of an island arc, today you'll find amphibolite and schist in place of andesites, basalts, and sediments. The distinctive Swakane terrane is exposed as gneiss, schist, and soaring roadcuts of gneiss and amphibolite along the highway between Wenatchee and Chelan. These rocks may be as old as 1.6 billion years.

The easternmost domain of the North Cascades is known as the Methow Domain. Here, two formations of very different ages provide an unmetamorphosed record of the ocean off North American shores. The oldest is the Hozomeen terrane. It consists of virtually unmetamorphosed cherts and pillow basalts that reveal ancient seafloor, 350 to 220 million years old (Mississippian to Triassic). These resistant rocks form higher topography east of Ross Lake, including Crater Mountain and Jack Mountain. Much younger sediments were deposited on top of this old seafloor. Known as the Methow terrane, these sandstones and shales were part of large 200- to 100-million-year-old submarine fans. The Methow terrane appears along the narrow, challenging road to Harts Pass near the east entrance of North Cascades National Park.

Mount Stuart (9,415 feet) is one of the most rugged and imposing granitic peaks of the North Cascades, although it lies well to the south of the National Park and only 20 miles north of Cle Elum, Washington. It is an outlier of the North Cascades. On the southern flanks of Mount Stuart, faithfully guarded by a large herd of big, wooly mountain goats, a thick slice of the seafloor is marooned at Ingalls Lake, and extends several miles to the south. Known as the Ingalls ophiolite, it includes mantle peridotite, gabbros, basalts, and sediments, along with some intrusive basaltic dikes. The Ingalls ophiolite is between 190 and 200 million years in age, is part of the Western Domain, and represents a chunk of the ocean crust, jumbled into North America by the accretion of terranes. Similarities in age, genesis, and rock compositions suggest that the Ingalls ophiolite was once part of the Josephine ophiolite in the Klamath Mountains, and was transported more than 250 miles northward by faulting.

In the Alpine Lakes wilderness, the peridotites of the Jurassic Ingalls ophiolite form orange-brown exposures, which contrast with the granitic rocks of the Cretaceous Mount Stuart batholith in the background.

The San Juan Islands echo the terranes of the North Cascades' Western Domain. These two areas of accreted terrane are likely continuous beneath the Puget Lowland. The most distinctive and accessible terrane in the San Juans is the Decatur terrane of Fidalgo Island—a mélange that includes a scrap of oceanic crust and upper mantle called the Fidalgo ophiolite. An ophiolite is a specific package of rocks that includes (bottom to top) ultramafic rocks, gabbro, basalt, and sedimentary rocks. This sequence is characteristic of oceanic crust worldwide, both today and in the past. Ophiolites usually mark the presence of a "convergent margin" where an oceanic plate collided with a continent.

About 150 to 170 million years ago most of the rocks on Fidalgo Island and adjacent islands floored the ocean west of North America. At Fidalgo Head and Washington Park, outcrops of gabbro and ultramafic (harzburgite and dunite) rock attest to the presence of deep-seated oceanic crust brought to the surface by faulting and collisions. In addition, the island's highest and most publicly accessible point, Mount Erie Park, is buttressed by diorite—the stuff of the magma chambers beneath island arc volcanoes. This diorite suggests that the Fidalgo ophiolite is a remnant of oceanic crust that supported volcanic islands like the modern Aleutians or Marianas. On Lopez Island, pillow basalts and ribbon chert, of the upper part of the sea-floor, dominate, and on Decatur Island pillow basalts and volcanic breccias are abundant. On the southern tip of Fidalgo Island, Deception Pass State Park provides excellent views of the Jurassic-Cretaceous sedimentary rocks at the top of the ophiolite. Together, all these rock units make up the Fidalgo ophiolite.

As North America inched westward, and the welter of ancient islands moved north and east, the stage was set for collision. In geology, collisions happen in (very) slow motion. Plates move at the rate a fingernail grows; collision and accretion require millions of years. For the Northwest, accretion was a complex process, involving sideswipes ("transpressional faulting") as well as head-on collisions. The dates for collision are imprecise. Based upon the dates of granitic rocks generated by the heat and pressure of the process, the islands (soon to be exotic terranes) first encountered North America in the Jurassic, about 130 million years ago. The process of adding and consolidating the new real estate likely continued until the Middle Cretaceous, about 90 million years ago.

The tan rocks exposed at Deception Pass are deformed and thickly bedded sandstones—the uppermost part of the Fidalgo Island ophiolite.

A broad band of heated, folded, and stretched rock exposed in western Idaho marks the boundary between the old continent and its newly acquired land. Known as the Salmon River Suture Zone, it is best exposed along US Hwy 195 near Riggins, Idaho. Here, rocks of the Wallowa island arc splinter, stretch, and intermingle with magmas generated as continent and islands met. Phyllites, slates, talc schists, and mylonites—all born of a moving and unhappy crust—are intertwined with mantle fragments and granite dikes. This boundary is also delineated by a change from oceanic to continental rock compositions, measured as the 87/86 strontium isotope ratio. West of the line, oceanic crust has a ratio of 0.704. East of the line, the ratio is 0.706. In a roundabout but precise way, this ratio gauges whether the deep crust is more calcium-rich, and hence oceanic (0.704), or whether the rock's progenitors are more potassium-rich, and hence continental (0.706). Using this measure, as well as geophysics, the boundary makes a sharp turn to the west just north of the Oregon-Washington line, and then fades from detection beneath much younger basalts.

Plesiosaurs, Dinosaurs, and Life on the Cretaceous Northwest Coast

The Northwest's first native coastline may have been oriented in a more east-west direction than today's familiar north-south. What we know today as the Columbia Basin likely took the form of a large oceanic embayment—a sort of ancient, larger Puget Sound. In terms of today's geography, the coastline extended from about Ashland, Oregon, through Dayville or Mitchell, northwest to Bellingham, Washington, and into British Columbia.

The Cretaceous was the heyday of dinosaurs. Although the fauna would definitely be different than we see today, its flora presented a world we might recognize. Flowering plants thrived. The ancestors of modern trees—willows, alders, magnolias, birch, and pines—provided shade, along with more ancient flora including cycads, palms, and massive 10-foot-diameter fern-trees (*Tempskya*). Their fossils are found not only near Mitchell, Oregon, but also 100 miles farther east, near the mining town of Greenhorn, where chunks of petrified *Tempskya* trunk are incorporated in gold-bearing Cretaceous or Early Tertiary (Paleocene) gravels. Grass was a new and very scarce invention. The climate of the Northwest emulated modern Georgia, Florida, or Louisiana. It would have been, as paleontologist Dave Taylor noted, "an exotic-looking area."

While tyrannosaurs and triceratops waged battles in Montana, Wyoming, Colorado, and Utah, the Northwest's landscape was placidly bereft of dinosaurs. There was little dry land for them to walk around on, except a scatter of distant, offshore islands, submerged plateaus, and a narrow coastline hemmed by mountains. Only one, relatively small, bone of a true dinosaur has been found or excavated anywhere in the Northwest. This skimpy find is a 3-foot-long portion of the backbone of a hadrosaur (also known as a duck-billed dinosaur), found in Cretaceous-age sandstones at Cape Sebastian on the southern Oregon coast. Hadrosaurs were vegetarians, grew to 20 feet or more in length, and probably, like giant, flat-billed, featherless (?) chickens, lived in groups (flocks), laid eggs in nests, and watched over their young. Paleontologist David Taylor, who found and excavated the fossil backbone in 1995, considered it to be the partial remains of an animal that died along a river close to the shore, and then was washed into the ocean.

This hadrosaur, like so many other Oregon residents, was not an Oregon native. It lived in southern California. The exotic Gold Beach terrane that is its home was faulted at least 600 miles north from its origin. The Cape

Cretaceous shales of the Hudspeth Formation near Mitchell, Oregon, revealed the skull of a plesiosaur, as well as fossils of mollusks, echinoderms, and sponges.

Sebastian hadrosaur may even be a relative of many other Late Cretaceous dinosaurs that are found in Baja, California.

What the Northwest lacked in dinosaurs, we made up in their cousins: toothy, ferocious marine reptiles and flying, reptilian pterosaurs. Pterosaurs are the parchment-winged personification of airborne ugliness, with a dash of redeeming acrobatic grace. Marine reptiles mimicked modern dolphins or the mythic Loch Ness Monster in shape and size, and came in three flavors: plesiosaurs, mosasaurs, and ichthyosaurs.

Plesiosaurs emerged as a Pacific Northwest specialty. They were apex predators of the Cretaceous seas, devouring fish and mollusks, as well as each other. Their colors were likely somber, and similar to modern mammal analogues of orcas and porpoise—we know that ichthyosaur skins contained black pigments. Ichthyosaurs and plesiosaurs gave birth to live young, and possibly nurtured their offspring in social family groups, according to researchers Robin O'Keefe and Luis Chiappe, who found that a plesiosaur specimen on display at the Los Angeles Museum of Natural History was a female with an embryonic baby in its belly.

In the Northwest, three major plesiosaur finds suggest that these animals frequented North America's Cretaceous shores. Near Mitchell, Oregon, the partial skull of a snaggle-toothed, thick-necked, seagoing hunter that resembled the imagined sea monster of ancient mariners was discovered in 2005. This 25-foot-long plesiosaur was an ambush predator that could swim at an

estimated 30 knots in short bursts. It was the tiger of the Cretaceous seas. Another plesiosaur—a long-necked elasmosaur, was found on Vancouver Island, British Columbia, in 1988 by 12-year-old Heather Trask and her father, Michael. This animal, which looked like the classic images of the Loch Ness Monster, was more likely a slow swimmer that traveled longer distances than the Mitchell plesiosaur. Then there is *Kourisodon puntledgensis*—a mosasaur (a marine reptile with a short neck, long tail, and very big teeth) that was about 15 feet long—also found in the Cretaceous rocks on Vancouver Island. Both Vancouver Island animals likely arrived in the Northwest pre-entombed as part of an exotic terrane. The same may be true of the Mitchell plesiosaur.

Oregon, Washington, and northern California have also yielded fossils of ichthyosaurs—porpoise-like marine reptiles that bore live young and had huge eyes that adapted them to dive to great depths to hunt fish and mollusks. They include a Triassic ichthyosaur, *Shastasaurus*, discovered in the Martin Bridge Limestone in the Wallowa Mountains, and another found in similar Triassic limestone at the base of Mount Shasta—hence the name. These animals arrived as part of exotic terranes that were accreted onto North America, after having lived their lives elsewhere.

Of course, the marine life in Cretaceous seas included smaller creatures. Mollusks, brachiopods, gastropods, and ammonites thrived off the Northwest's new coast. Some of the "smaller" creatures were quite large—including 3-foot-diameter coiled ammonites.

Eagle Cap rises above Mirror Lake in northeastern Oregon's Wallowa Mountains. Glaciers crafted the landscape here, exposing the Jurassic to Cretaceous granitic rocks of the Wallowa batholith.

Gneiss and Stitching Plutons: Stabilizing the New Land

As the collision of North America and exotic terranes began, the heat and pressure of collision melted the most deeply buried of the accreted terranes. Rocks that were volcanic islands, scummy seafloor, and bright limey beaches transformed into thick, molten granitic slush (about 1,800 degrees F). These now-liquid mixtures rose and melted their way through the accreted rocks above them. Today their solidified remains are the white granitic rocks that adorn some of the Northwest's craggiest peaks. They are the "granites" of the Sierras, the Wallowas, the Elkhorn Mountains and Anthony Lakes, of the Seven Devils; the Klamath's Mount Ashland and Grayback pluton; the banded gneisses of the North Cascades. Even the Idaho batholith owes its genesis, 90 million years ago, to the heat and pressure of collision.

Although we commonly call these light-colored, intrusive, igneous rocks "granite," they are chemically and mineralogically different. True granites contain at least 20 percent quartz and 10 percent potassium-rich feldspars. The "granites" of the Northwest include less of these minerals. Most are granite look-alikes with more iron and calcium, reflecting their origins from altered seafloor. Geologists classify most as the granitic rocks granodiorite or tonalite.

The intrusive bodies of granitic rocks are known as stitching plutons. They were viscous magmas that were sometimes emplaced with force, often following the weak seams of terrane boundaries. As a result, they glued or "stitched" two different terranes together. Not all the Jurassic-Cretaceous plutons follow this rule, but most do, so that the idea and the name "stitching pluton" stuck.

In the Klamath Mountains, the stitching plutons were produced in three different generations. Most Klamath granitic plutons span the time from 174 to 136 million years ago. They include the Wooley Creek batholith (162 Ma), Ashland pluton (152 to 161 Ma), Grants Pass pluton (139 Ma), Craggy Peak (138 Ma), and Castle Crags (130 to 140 Ma). Their trace-element compositions indicate that all were generated by melting of older metamorphosed volcanic and other igneous rocks, with variable proportions of sedimentary rocks. Studies of tiny zircons in the Klamath granitic rocks indicate that some of the buried rocks that melted were Precambrian in age, and perhaps as old as 1.7 billion years. But whether these were derived from the North American continent or from a remnant microcontinent somewhere in the Panthalassa Sea (today's Pacific Ocean) is unknown.

In the Blue Mountains, plutons include two large bodies: the Bald Mountain batholith in the Elkhorn Mountains (about 148 to 157 Ma), and the Wallowa batholith (140 to 125 Ma), according to dates by Joshua Schwartz and others. But many smaller intrusions are present as well, including small, mafic (iron-rich) or sodium-rich intrusions, 160 to 150 Ma, that flank the Wallowa batholith. These include the Cornucopia Stock just north of the town of Halfway. To the east, the Seven Devils are a smaller mirror image of the Wallowas. To the west, plutons exposed in the Greenhorn Mountains are part of the same generation, as well as rocks south of Unity in Bull Run Ridge and west of the Snake River on Lookout Mountain.

In the North Cascades, the stitching plutons were deformed and metamorphosed and look more like metamorphic rocks than granites. They are represented by the Skagit Gneiss Complex, including the rocks of Mount Misch, Sulphur Mountain, and Eldorado Peak. Still, these now-gneissic rocks are relatively young—70 to 90 million years, or Late Cretaceous in age. Their banding was acquired during a later, Eocene, episode of crustal stretching, and is generally not an original feature of the rock. The Mount Stuart batholith (96 to 91 Ma), which occupies most of the Alpine Lakes Wilderness including the Enchantment Lakes Basin, is among the largest and best exposed of Washington's stitching plutons, covering about 270 square miles. The granodiorites and quartz diorites here lack the banding of the Skagit Gneiss, although they are similar in age. The oldest rocks in the Mount Stuart batholith (96.3 to 95.4 Ma) occur in the northwest part of the batholith near Icicle Creek; the youngest (90.9 to 90.7 Ma) occur near the south margin with the Ingalls ophiolite.

Rhyolites and silica-rich tuffs (above) are interbedded
with fossiliferous sandstones near Denning Spring, south of
Pendleton, Oregon.

CHAPTER 6   Paradise Found   *Life in the Tropics*

The first geologic epoch of the new Cenozoic era is the Paleocene—the "Old New Age." It began on the day when the last dinosaur died, 66.038 ± 0.025/0.049 million years ago, and lasted until a brief episode of intense global warming (between 55.728 and 55.964 Ma) erased most marsupials and about 40 percent of marine life. During the Paleocene's approximately 10-million-year run, mammals emerged from their burrows and realized that they were now the best and biggest show in town. They experimented. Marsupial mammals—ancestors of opossums and kangaroos—rose to prominence. Some, like *Titanoides primaevus,* reached 6 feet in height. Their sharp, piercing teeth, similar to modern opossums, suggest they were—at least in part—carnivores. But pantodonts were far outsized by other Paleocene experiments. There was *Uintatherium*—an elephant-sized, rhino-like animal weighing an estimated 3 tons and sporting multiple horns on its

face—as well as 6-inch saber teeth in its upper jaw. This browser of Utah shrubs and trees was among the largest land mammals ever.

The Paleocene marks a relatively warm time in Earth's history. There were no polar ice caps. Near Pendleton, Oregon, some of the Northwest's first volcanoes erupted ash and gooey rhyolite lava. Palms and avocados flourished, along with ferns, willow, and alder. Bald cypress grew in marshy, tidal bayous. By this time, Pendleton was close to its present position about 45 degrees north, so we cannot blame the nearly subtropical climate on a more southerly position.

Carbon dioxide concentrations were likely about 600 to 500 ppm on average (versus today's 400 ppm), based on plant stomata and other data. Equatorial latitudes were 10 to 12 degrees F warmer than today. Polar regions were as much as 40 degrees F warmer, according to M. Heinemann and colleagues at the Max Planck Research School for Climate Modeling.

The location of the Pacific Northwest's Paleocene coastline is uncertain. The best guess suggests the beach lay on the east side of today's High Cascades (which at the time were far in the future) with an embayment that swung from Portland east to Pendleton, and then looped westward to Republic and Bellingham.

Basalts erupted not far offshore. Today, these rocks compose the coast ranges from Roseburg to Corvallis, north to the Willapa Hills and farther into today's Olympic Mountains. They developed near the Paleocene-Eocene shore and seamounts and a submarine plateau. Sandstones, shed from the continent and deposited as deep-sea fans, are interbedded with the basalts and pillow lavas. These rocks are known variously as the Umpqua Formation, the Looking Glass Formation, and the Siletz River volcanics. The oldest rocks are about 65 million years in age, deposited just after the dinosaur-eradicating cataclysm.

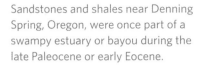

Sandstones and shales near Denning Spring, Oregon, were once part of a swampy estuary or bayou during the late Paleocene or early Eocene.

Tropical leaves up to 15 inches long are part of the warm-temperate Denning Spring flora.

Eocene pillow basalts inhabit Mary's Peak, the highest summit in Oregon's Coast range. Though about 56 million years in age, these pillows still display a glassy rim and radial joints typical of pillow lavas.

The oldest of these seaworthy pillow basalts (58 Ma) rear their lumpy heads in outcrops on the campus of Umpqua Community College in Roseburg, and in roadcuts throughout the area. Northward along Oregon's Coast Range, similar rocks yield progressively younger ages. The basalts of Mary's Peak are Eocene, 52 million years old. Willapa Hills basalts clock in at 50 million. On the Olympic Peninsula, the gnarled outcrops of Mount Zion, Stripped Peak, and Bremerton Hills also rank at about 50 million years in age, but the Crescent Basalt—pillow lavas that festoon the eastern Olympic Mountains—date to 48 to 45 million years. Ultimately, all these oceanic basalts and related sediments were accreted to North America by 45 million years ago. This terrane, known as Siletzia, constitutes the last exotic piece of real estate added to the Pacific Northwest.

In the North Cascades, the heat and pressure of this collision produced multiple small, shallow intrusions that date from about 68 to 55 million years in age, including the Oval Peak batholith (61 Ma). Collision and subsequent crustal stretching in the North Cascades and points east also reformatted some stitching plutons into gneissic, metamorphic rocks. These include a gneiss exposed near Lake Juanita, and the Diablo orthogneiss—easily accessible and exposed at the Diablo overlook on the North Cascades Highway (WA Hwy SR 20). Another larger North Cascades pluton was transformed into the pervasive Skagit Gneiss that is well exposed in roadcuts at the Diablo overlook site, and other locations along

Hwy 2. These rocks have all the blazing white and black mineralogy of their parents, along with some compelling aesthetic bands and other patterns.

The Paleocene–Eocene Thermal Maximum

Today we are rightly concerned about global warming driven by greenhouse gas. The dramatic finale of the Paleocene period (along with prior extinction events) demonstrates that geologically sudden greenhouse gas emissions drive the planet to climatic extremes. A worldwide spike in temperatures occurred between 55.728 and 55.964 million years ago, based on high-precision dating by Adam J. Charles and others at the University of Southampton. Known as the Paleocene–Eocene Thermal Maximum (PETM), the global heat wave lasted only about 100,000 years. Some scientists estimate its most severe high temperatures persisted for only 20,000 years. Recovery from the thermal climax (55.832 Ma) took another 50,00 to 70,000 years.

The effects were hauntingly similar to what we are observing today. Globally, oceans warmed and became more acidic, while temperatures on land increased markedly. Polar regions felt the brunt. Arctic Ocean temperatures were a balmy 76 degrees F at the height of the PETM and would remain elevated until about 38 million years ago, according to Kenneth Rose of Johns Hopkins University.

Precise dates of this event indicate that the onset of the PETM did *not* occur at the peak of a 400,000-year cycle in Earth's orbital eccentricity. Instead, it occurred on the falling limb of a cycle when warming by the sun was decreasing. This strongly suggests the culprit was greenhouse gas rather than solar influx. There are two indicted suspects for the source of the greenhouse gasses. First, huge eruptions in the North Sea belched CO_2 and built a vast submarine basaltic plateau that today pokes it head above water as the Faroe Islands. Second, and more importantly, increased heat on the seafloor near these eruptions initially warmed the seafloor, releasing huge amounts of methane.

Atmospheric CO_2 reached about 1,200 ppm for a brief period of about 1,000 years. But the real villain during the PETM was methane—a gas with twenty-three times the greenhouse power of CO_2. Many Paleocene researchers have evidence that as oceans warmed, methane stored in a solid state (methane clathrates) changed to methane gas, rose through the water column, and entered the atmosphere. Today, methane is measured at 0.7 parts per million in the atmosphere. During the warmest of the PETM, methane reached more than double that level, or 1.6 ppm. This may seem insignificant, but this change was sufficient to raise average global temperatures 12 degrees F worldwide and send polar temperatures into overdrive. At the Paleocene thermal maximum, the temperature of polar ocean waters exceeded 75 degrees F. On an average summer

day at the North Pole, the thermometer hit 100 degrees F.

The consequences of this brief greenhouse gas binge included major marine extinctions. Warming of deep waters and subsequent oxygen deficiency wiped out about half of the deep-sea foraminifera. The number of plankton species and diversity decreased dramatically. However, vast planktonic blooms were likely one mechanism that pulled excess carbon from the atmosphere and sequestered it on the seafloor as part of the cooling-off process. Carbon isotopes from the Bighorn Basin, Wyoming, also suggest that land plants picked up their rates of photosynthesis and growth, which led to more rapid weathering and rates of soil formation—and additional carbon sequestration. Ultimately, notes Paul Koch and colleagues of the University of California, Santa Barbara, higher atmospheric CO_2 led to higher rates of oxygen production. This, ultimately, lowered global temperatures and marked the end of the high-CO_2 high-temperature event.

On land, there was a new charismatic megafauna. Marsupials declined precipitously. But this vacuum was quickly filled with the rise of placental mammals. As the global thermometer cooled, two new orders of placental mammals seemed to arise almost overnight. The perissodactyls, or hoofed animals with one or three toes, included ancestral horses, more reasonably sized rhinos (*Teleocerus* and brontotheres), and tapirs. The artiodactyls, or even-toed hoofed animals, which appeared between 55 and 53 million years ago, included ancestral pigs, camels, deer, hippos, and surprisingly, whales. North America appears to be the hotspot for many of these origins.

This global cooling ushered in a new time period, the Eocene, when life would look more familiar.

The Eocene: Tropical Climates Continue

The term Eocene comes from the Greek *eos* (dawn) and *kainos* (new). Its beginning, calibrated at 54.9 million years ago, coincides with the appearance of the horses/tapirs (perissodactyls) and cloven-hoofed animals (artiodactyls). Neither of these evolutionary events is marked by a moment in time, the simple tick of a watch, or the midnight birth of the first *Hyracotherium* (three-toed horse) foal. However, the appearance of these two major mammal orders was geologically abrupt, occurring in as little as 100,000 years.

Eocene fig leaf, Clarno formation. The middle Eocene was a time of tropical splendor in today's central Oregon, with a forest of figs, palms, early bananas, and ancestral coffee trees.

Throughout the Eocene, the globe remained warm. Crocodiles basked on Greenland's shores, and a python the size of a school bus prowled the Amazon. It was a time when mammals would take the global helm, and forests, worldwide, were luxuriant and tropical.

The lands of today's Pacific Northwest lay close to their present latitudes, at 40 to 45 degrees north. But the forests were a subtropical mixture of palm trees, magnolias, and figs, along with more temperate species, including maples, oaks, and pines. The globe was warm, and the temperature difference between polar and equatorial temperatures was much less than today.

The Eocene atmosphere held more carbon dioxide than today's atmosphere. Estimates, based on carbon isotopes, oxygen isotopes, leaf form, leaf stomata, and marine sedimentary records, vary widely, from about 560 ppm to as much as 2,000 ppm CO_2. Whatever this experimental disparity, scientists agree that greenhouse gases played a significant role in maintaining global temperatures 6 to 10 degrees F higher than today.

It was a new age of tectonics and volcanoes as well. Subduction zones shifted. Mid-ocean ridges rotated. As a result, the Eocene saw diverse volcanism across the Northwest and neighboring Interior West, including volcanoes spouting basalts in Montana and central Idaho. Intrusion and crustal extension plagued the Okanogan region. As fascinating and readily apparent as the Eocene's lush, tropical ecosystems are, its tectonics remain shrouded in mystery.

Volcanoes Continue Construction of the Northwest

Cascade-like composite volcanoes, as high and as extensive as Mount Hood, inhabited eastern Oregon's John Day Basin 53 to 45 million years ago. Known as the Clarno Formation, these rocks were once considered subduction-related volcanoes similar to the Cascades. Andesite is their most abundant type of volcanic rock. This would seem to link the Clarno to conventional subduction processes. But many Clarno basalts, including Mitchell Butte and Marshall Butte, have chemical signatures that suggest an origin in the very deep mantle rather than at the more shallow levels associated with subduction. Other basalts contain olivine along with exotic quartz crystals (xenocrysts). Clarno rhyolites and dacites—silica-rich lavas that form Keyes Butte and other intrusives—also display trace element compositions that fit better with rifting than subduction.

The remnants of volcanic flows and domes and necks formed buttes near Mitchell, Oregon—White Butte, Black Butte, Amine Peak, Sand Springs Butte, Byrd Rock, and Mitchell Butte. Other peaks implicated in Clarno volcanism include Lookout Mountain and Round Mountain east of Prineville, and Ragged Rocks north of Prairie City. Along Cherry Creek, about 25 miles northwest of Mitchell, Oregon, the extraordinary basaltic dikes and basaltic lava flows (known as Good Rocks, for the Good family that settled there in the late 1800s) suggest a rift or extension

North of Mitchell, Oregon, the remnants of Eocene volcanoes and dikes make for a rugged landscape.

Good Rocks are the remains of an extensive system of basaltic dikes that fed Clarno volcanic eruptions.

zone in this ancient volcanic chain. Elsewhere, including Keyes Butte, light-colored dacites built volcanic domes.

Where there are composite volcanoes, there are mudflows. In this jagged Clarno landscape, ancient lahars (volcanic mudflows) are exposed as the spectacular cliffs of the John Day Fossil Beds National Monument's Clarno Palisades, south of Fossil, Oregon. Today, this is a dry landscape of sagebrush and juniper, but 50 million years ago it was a subtropical paradise. Fossils found in the Clarno Nut Beds include small bananas (*Ensete*), palm trees, dates, fig trees, ancestral coffee and tea plants, walnuts, and early grapes. More temperate deciduous species are present as well, including oak, elm, and maple. The mudflows also preserved the upright trunk of a katsura tree (the Hancock Tree), which today is native only to regions of Japan with mild climates. Altogether, 175 plant species flourished in the Clarno's diverse forests.

The Clarno Formation also provides one of the globe's most complete records of Middle Eocene (40 Ma) mammals. At the renowned Mammal Quarry, now protected as part of the John Day Fossil Beds National Monument, the vine-laden forests were habitat for a diverse fauna. Animals that ran, strolled, ate, and slept here included *Haplohippus*, an early horse that was a four-toed, dog-sized browser of soft tropical leaves, and a slightly larger three-toed horse (*Orohippus*) from the Late Eocene. Brontotheres, distant relatives of today's rhinos, also roamed the thickly forested volcanic landscape, along with *Achaenodon*—a bear-sized omnivorous pig. The apex predators of the Eocene likely included *Hemipsalodon,* a bear-shaped creodont that was about 25 percent larger than the average modern polar bear. It may have been mostly a scavenger, based on teeth and jaw structure, but its meals likely included many of the herbivores found in the mammal quarry, including early tapirs, *Haplohippus, Epihippus* (small, three-toed horses), *Teleoceras* (pig-sized rhinos), brontotheres, and *Heptacodon* (a slim relative of early deer). *Patriofelis*—a cat-like predator about the size of a modern cougar but more compactly built—also prowled. Most of this fauna, or similar species, are extant in Eocene localities across North America. And it's likely that most of them romped through the Northwest, including the lakeside forests of Republic and lush wetlands of Bellingham, Chehalis, and Roslyn.

Volcanoes appeared in Idaho and Washington as well. Volcanic and intrusive rocks cover almost 10,000 square miles in central and eastern Idaho. They are named the Challis volcanics for the town of Challis, Idaho. Related volcanic and intrusive rocks form the Absaroka Range and Crazy Mountains in Montana. The Challis volcanics seem to have little chemical linkage to a subduction zone. Instead, they are more closely related to crustal stretching, and extension.

In north-central Washington, the Colville igneous complex—a 1.6-mile-thick layer-cake of igneous and volcaniclastic sedimentary rocks—represents Eocene volcanic and intrusive activity. It is composed of the Colville batholith (granitic rocks exposed on the Colville Reservation), the San Poil volcanics, and the Klondike Mountain Formation (a sequence of basalts, andesites and rhyolites with abundant tuffs and fine-grained lakebeds near Republic). Near the small town of Republic, rose-colored cliffs of rhyolite, dacite, andesite, and tuff frame the San Poil River's narrow canyon. They were lava flows and hot, fast-moving tuffs 52 to 51 million years ago. Like the Challis Group, they originated from stretching and melting of the accretion-thickened crust. The process is called "delamination," and includes the decompression melting of crust and mantle as the crust—overthickened by accretion and folding—collapsed back into the mantle. The San Poil volcanics delineate the Republic graben—a faulted valley with the small towns of Republic at its head and Roosevelt on its southern end—that

Volcanic debris flows, known as lahars, form pinnacles on this un-named butte as well as the nearby Clarno Unit of the John Day Fossil Beds.

Rhyolite ash-flow tuffs and andesite lava flows make up much of the San Poil volcanics exposed in the San Poil valley on the Colville Reservation.

The plant seeds at Republic—fruity, edible, and easily transported—provide an additional clue to life in this extraordinarily diverse forest. Think apples, chestnuts, and figs, as well as currants, rosehips, elderberries, wild cherries, blackberries, raspberries, and others. All are menu favorites for birds. And the discovery of at least two well-preserved fossil feathers in the lakebed sediments suggest that birds, both large and small, occupied this forest.

Sparrows, swallows, woodpeckers, parrots, and robins would have found abundant food. The lakebed's fine sediment also entombed and recorded a diverse insect fauna. They range from the cute (aphids and ants) to the creepy (cockroaches and stinkbugs). In between are lacewings, staghorn beetles, pirate bugs, water-boatmen, caddisflies, mayflies, damselflies, and dragonflies. Bees, wasps, butterflies, moths, and many other insects lived and died around the lake.

At least six kinds of fish lived in this large lake. They include ancestral bowfin that grew

A well-preserved dragonfly found in the Stonerose Fossil Beds indicates that dragonflies of 45 million years ago were similar to today's.

marks the eastern boundary of this extended, delaminated crust.

Life in the Eocene Forest at Republic

Some 48 million years ago, the place that would become Republic, Washington, lay in the middle of a lake shaded by oaks and metasequoias. You can explore the ancient lake and its biology at the Stonerose Interpretive Center in downtown Republic.

This sizable lake stretched at least from Republic north to Curlew and then about 10 miles west. Its quiet waters collected ash-laden sediments and served as repository for the cast-off leaves and branches of regional vegetation. Plants that abandoned their leaves in

this lake included hawthorns, metasequoias, ginkgoes, early oaks, and alder. Growing near the lake was the signature flower of Stonerose—actually not a rose at all, but a plant called *Florissantia quilchenensis*—a close relative of modern cocoa (chocolate) and mallow plants. Other plants documented in the extraordinary lakebeds at Republic include ancestral forms of sassafras, plane tree, sycamore, maple, clematis, currant, katsura, Asian wheel tree, linden, fig, beech, horse chestnut, elm, sumac, hydrangea, mock-orange, and apple. The Eocene forest at Republic had at least three species of pine trees—and the richest concentration of fossil conifer genera anywhere in the world, according to the late paleobotanist Wes Wehr of the Burke Museum, University of Washington.

to 15 inches long, as well as ancestral mooneye, goldeye, minnows, and suckers. There was also an ancestral salmonid (*Eosalmono driftwoodenis*), a fish poised midway in its evolutionary journey between grayling and salmon. Crayfish prowled the lake's rocky shoreline.

The Great Pacific Northwest Coal Swamps

The fossilized remains of forests that grew in very moist conditions 55 to about 45 million years ago are sequestered on the east and west sides of the Cascades. Today, they are (or were) coal deposits. During the last two centuries, we have been busy putting a lot of their sequestered carbon back into the atmosphere.

Along the shores of Puget Sound near Bellingham, the 3-mile-thick Chuckanut Formation contains an astounding diversity of fossilized Eocene plants. This was a lush, multistoried subtropical forest. Today, much of the forest has been compressed to mostly bituminous coal beds up to 15 feet thick. More than 25 million tons of coal was mined from the Chuckanut Formation, beginning in 1883. The Chuckanut, which is well exposed along Chuckanut Drive (Washington Hwy 11) south of Bellingham, is renowned for its fossils of radiating palm fronds (*Sabalites*). The diverse forests included many other plants and trees, including giant horsetail rush, sassafras, swamp cypress, water pine, sycamore, alder, birch, and hydrangea. Based on leaf forms and sizes, the mean annual temperature was 59 degrees F, with a mean winter temperature a balmy 50 degrees F. Today, Bellingham's

average annual temperature is a cooler 50.3 degrees F, and average low temperature is 42.7 degrees F.

To the south, similar forests produced coal deposits near Centralia, Washington, which were mined from 1833 until 2006. Coal beds were as thick as 40 feet, and annual production of coal in the Centralia mines averaged 4.1 million tons annually.

The Eocene landscape on what is now the east side of the Cascade Mountains (but then, with no Cascades, it was likely a continuous plain) was a relatively flat, meandering river system replete with braided streams, backwaters, meanders, oxbow lakes, and bogs. Episodic floods swept down from the highlands around Mount Stuart, spreading gravels across the swampy flood plains. The result included the Roslyn and Taneum-Manastash coal deposits—mined from 1886 until 1963, producing as much as 2 million tons of soft bituminous coal per year.

The fertile swamps and lush tropical forests of Bellingham, Chehalis, and Roslyn were contiguous during the Early Eocene. However, faulting along the Straight Creek Fault system (including the Darrington-Devils Fault) began pulling these forests apart and isolating them as separate basins beginning about 50 million years ago. By the time climates shifted and the tropical forests faded about 35 million years ago, these basins were likely discrete but broad valleys, separated by 80 miles of mountainous and increasingly volcanic terrain. Today, the Straight Creek Fault also defines some

significant offsets in the North Cascades, but seems to be completely inactive.

This flood-prone subtropical forest must have been a very, very healthy place to live. Evidently, no animals died here—with the single exception of a turtle, there are no known animal fossils from either the Chuckanut or Swauk Formations. Fortunately, it was muddy. In lieu of bodies, animals left tracks. Paleontological sleuths have identified the prints of the 20-inch-tall dawn horse (*Hyracotherium*), or possibly a tapir, as well as "large, short-legged animals" thought to be the last of the large pantodonts, or perhaps first of the hippopotami, who appeared on the global stage about 54 million years ago. Tracks of a cat-like creodont, and of *Diatryma giganteus,* a 6-foot-tall, omnivorous, 385-pound flightless bird (think ostrich with a hawk's beak, on steroids) were also found in the Chuckanut in 2010.

The Eocene Shore

Organic detritus from the vast Eocene coastal swamps gathered in deltas and offshore fans. Plankton and kelp lived and died in these waters. These sediments sequestered sufficient organic material to produce natural gas, which was first tapped in 1979 as the Mist Natural Gas Field, located near the town of Mist in Oregon's northern Coast Range about 60 miles northwest of Portland.

To the south, similar coastal swamps and estuaries of the 44-million-year-old Coaledo Formation also produced coal. These deposits of sub-bituminous lignite were mined from

1854 to the late 1930s, with about 3 million tons of coal commercially extracted. Marine fossils occur in Coaledo coastal deposits. These include sand dollars, brittle stars and other starfish, mollusks, gastropods (such as *Turritella*, a tightly coiled and pointed shell), and the teeth of both sharks and rays. Thomas Condon, Oregon's first state geologist, categorized the age and marine nature of the sandstones in the late 1880s when he recognized that fossil scallops called *Venericardia*

in rocks near Charleston, Oregon, were the same Eocene fossils found near Paris, France. Accurate dates of 40 to 44 million years for the French fossils corroborate Condon's conclusion.

The warmth of the Eocene ocean is evident from the shellfish that lived in Pacific Northwest waters. Native shellfish were thermophillic (warmth loving). They included *Xenophora*, with a small shell that looked like a turban; *Eocernina*, with a small,

narrow, tightly spiraled shell; and *Lyria*, with a conch-like shell, all found in the Humptulips Formation of coastal Washington. Crabs, found near Wahkiakum, Washington, included the small but heavily clawed *Portunites*. In the sandstones and conglomerates of the Cowlitz Formation of southwest Washington and northwest Oregon, marine animals that include corals and nautiloids represent a warm, tropical to subtropical, shallow marine environment—probably a river delta and related bays. The Mississippi Delta, and nearby Louisiana bayous, might be an apt analogy, although the Coaledo Formation is bereft of "gators" and the paleobotany of the Coaledo forests has largely vanished with their coal.

The Eocene Crust Extends and Rotates

As North America began colliding with the rocks of exotic terranes about 120 to 90 million years ago, it also overran the subduction zone and down-going slab of the now-vanished Farallon plate that once formed the Pacific's crust. By about 60 million years ago, as North America moved west, the old subduction zone served as a conduit for mantle heat to rise beneath the continent's crust, buoying, stretching, and melting the rocks that now are part of the Okanogans. The crust, especially in eastern Washington, began stretching.

The subsurface of the Okanogan area, from Colville and Republic to Kettle Falls, began pulling apart. The strongly banded gneiss of Stowe Mountain is an example of

The Eocene Coaledo Formation in southwest Oregon included coal deposits and plant fossils. At Cape Arago State Park, the sandstones include odd concretions and gentle folds.

The banded rocks found in the Kettle Hills are part of the Kettle dome—a structure caused by the stretching and extension of northern Washington's crust during the Eocene.

deeply buried sediments that were stretched and stressed so much that they melted. The stretching and melting lasted at least 8 million years, from 60 to at least 52 million years ago, according to Seth Kruckenberg and his colleagues at the University of Minnesota. Today, the rocks near Colville resemble ordinary granites and ordinary gneiss. But they are not ordinary intrusions. They are melted lower crust, which domed upward, buoyed by its own heat and low density, and stretched like a giant taffy-pull. Most display a banded fabric generated by heat and extension. They are known as "mantled gneiss domes." The rhyolites and related rocks of the San Poil volcanics and the Republic Graben are additional consequences of this Paleocene to Eocene crustal heating, melting, and stretching.

To the west, the newly acquired North Cascade terranes also felt the stress of new subduction and crustal stretching. The more deeply buried portions of the Methow terrane melted, producing the true granites of the Golden Horn batholith. Its stunning scenery adorns the east entrance of North Cascades National Park, where the imposing Liberty Bell and Early Winter Spires are sculpted from the 49-million-year-old Golden Horn batholith.

Golden Horn batholith, exposed over about 120 square miles of the North Cascades, is a large and somewhat strange granitic intrusion. It was generated by melting the seafloor sandstones and other sedimentary rocks of the Methow terrane, and cooled at geologically shallow depths of about a mile. This near-surface emplacement allowed the development of vugs, which are mineral-rich cavities. Its source rock—the Methow terrane's continentally derived sediments—provided large quantities of sodium and potassium, as well as fluorine, chlorine, zirconium, and many other trace elements. This translates into a true "A-type" granite that contains abundant quartz and orthoclase feldspars and some unusual minerals, including faintly blue-colored "hornblende" (technically, the minerals riebeckite and aenigmatite). The vugs, also known as miarolitic cavities, are a mineral collector's paradise, with fluorplumbopyrochlore, calciocatapleiite, and at least fifty other rare minerals.

Central Washington was also home to smaller and darker Eocene intrusions spawned by crustal extension. The Teanaway dike swarm, exposed near the confluence of the Teanaway and Yakima Rivers, is about 47 million years in age. (Some whole rock dates are as young as 39 million years.) It is the third largest basalt province in Washington, behind the Columbia River basalts and the Crescent basalts/Siletzia found in the Olympic Mountains and Coast Ranges. Most of the Teanaway's volume has been eroded away, but the conduits to eruptions—basaltic dikes—are abundant. The Teanaway basalts were likely generated by melting of altered rocks (in this case, the rocks of the lower lithosphere, including gabbro and peridotite) as the crust stretched and heated. Teanaway dikes are nicely exposed in roadcuts north of Cle Elm and along US Hwy 97. Their WNW orientation indicates that in the Middle Eocene, the crust was indeed stretching—in a southwest to northeast direction.

The Golden Horn batholith is an Eocene pluton that solidified within about 3 kilometers of the surface. It is well exposed in Early Winters Spires and Liberty Bell near the east entrance to North Cascades National Park.

Oregon's Painted Hills, John Day Fossil Beds National Monument, record the change from the warm, forest-rich landscapes of the Eocene to the temperate grasslands of the Oligocene in their ancient soils. Red bands are tropical soils, lighter colors indicate cooler, drier climates. Upward through the sequence, the soils become progressively lighter, the climates cooler and drier.

Global Temperatures

Millions of years | 2500 | PROTEROZOIC | 541 | PALEOZOIC | 252 | MESOZOIC | 66.0 | CENOZOIC

Camb | Ordo | Silur | Dev | Miss | Penn | Perm | Trias | Juras | Creta | Paleo | Eoce | Olig | Mioc | Plioc | Pleist | Holo | Anth

CHAPTER 7   Grasslands Take the Stage   *Calderas and Cooling*

While the Eocene is renowned for its global warmth, the time period that follows is known for global cooling. The Oligocene, 33.9 to 23.03 million years ago, began with an abrupt cooling event that lasted for about 350,000 years, with most of the cooling occurring in the last 40,000 years of that period. This shift triggered the temperate world in which we live today. Coastal coal swamps vanished. Jungles faded.

In central Oregon, vegetation changed from viney subtropical forests to deciduous woodlands similar to the native vegetation of today's Willamette Valley—maples, oaks, sycamores, and metasequoia. (Metasequoia, which was extinct in the Pacific Northwest for about 20 million years until discovered in China and returned to its native homeland, now grows in Portland as a 10-foot-tall hedge at McMenamins Kennedy School; a tree on the campus of Whitman College in Walla Walla, Washington; and a grove at their ancient home at the John Day Fossil Beds Painted

Hills unit, among many places.) The shifts are well documented in many fossil sites, including the John Day Fossil Beds and fossils near Goshen and Comstock, Oregon.

This climate change was induced by plate tectonics, in collaboration with volcanoes, ocean currents, and biology. The transition to truly temperate mid-latitudes and polar ice caps took 3 to 6 million years. The culprits in this global cooling included plate motions—and plants' ability to suck CO_2 out of the atmosphere.

Until about 45 million years ago, Antarctica was tethered to South America. That connection broke slowly as Antarctica rotated and separated from Australia, creating the Drake Passage about 40 million years ago. As the passage widened and deepened, cooler water began to flow from Atlantic to Pacific, ultimately creating the Antarctic Circumpolar Current (ACC). What was at first a gradual process accelerated between 35 and 34 million years ago. Isolated from warmer currents from

the Atlantic and Pacific, the seas of Antarctica cooled. And eventually, so did the land. Antarctica's first snows fell 33.7 million years ago. According to dates on ice cores recovered in 2013, the continent bore substantial glaciers by 33.6 million years ago.

There was more to global cooling than isolating and chilling the Antarctic, or removing the vegetation. The compositions of both the atmosphere and the oceans changed as well. In a sort of climatic Rube Goldberg mechanism, they are linked to one another, and both are related to the shift in oceanic circulation. Essentially, more snow on land led to falling sea level. Lower sea levels resulted in smaller submerged areas of continental shelves erosion, and burial of carbonate formerly stored there in short cycles. At the same time, A. J. P. Houben notes, the new, vigorous circulation in the eastern Pacific created stronger planktonic blooms of more robust plankton, which sank and sequestered even more carbon on the seafloor. The records in deep-sea sediments show that as the planet cooled, atmospheric CO_2 dropped, and carbon storage in the oceanic sediments increased. And although not a straight-line progression, that trend continued from 33.7 million years ago until about the year 1820, when humans began a rapid reversal of global carbon storage.

The changing vegetation also provided climatic inputs according to the University of Oregon's Greg Retallack. In mid-latitudes, jungles, with dark, heat-absorbing vegetation and water-transpiring talents that maintained leaf-friendly levels of humidity, shrank. Open savannas and grasslands replaced deciduous forests. The planet's albedo—the ability to reflect light—and heat—back into space increased. And temperatures dropped.

Brian Jicha of the University of Wisconsin has suggested that explosive eruptions of the American Southwest's ignimbrite flare-up, and other powerful volcanoes, might have contributed to this cooling trend. He documented more than 500 eruptions of considerable volume that occurred between 35 and 28 million years ago, which corresponds to a minimum frequency of one major eruption about every 13,000 years. This activity included some Pacific Northwest calderas that are only now being mapped. Powerful caldera eruptions would inject huge volumes of both ash and sulfur high into the atmosphere. Once there, SO_2 reacts with OH and H_2O to form H_2SO_4 aerosols. These sulfate aerosol droplets reflect and/or absorb solar radiation, which reduces the amount of solar energy that reaches Earth's surface. This produces global cooling.

Meteorite impact might also have triggered—but not sustained—global cooling. Candidates (all dated at about 35 million years in age) include the meterorites that caused the 53-mile-wide Chesapeake Bay Impact Crater on the southern end of Delaware's Delmarva Peninsula, Toms Crater on the continental shelf about 100 miles east of Atlantic City, and the 60-mile-wide Popigai Crater in Siberia.

Work by Greg Retallack of the University of Oregon, and work published in 2013 by Timothy Gallagher and Nathan Sheldon of the University of Michigan, finds that on the east side of Oregon and Washington, average surface temperatures fell almost 5 degrees F between 34.3 and 33.3 million years ago. Oceans cooled 5 to 8 degrees F in the same time interval. Although this is a snail's pace compared with anthropomorphic climate warming, it is rapid for geologic systems.

Oligocene Calderas in the Northwest

The Oligocene was perhaps the most bombastic time in the geologic history of the American West. Huge explosive volcanoes, known as calderas, erupted torrid ash flows in Nevada, Colorado, New Mexico, Arizona, and Mexico. And Oregon. This geologic temper tantrum is known as the Oligocene ignimbrite flare-up. Its cause is uncertain.

Calderas are deceptive volcanoes. Their topography is subtle. Instead of building mountains of lava, ash, and mudflows, they blast most of their erupted material into the

The tawny tuffs of Smith Rock are the northern ramparts of the Crooked River caldera—a 25- by 20-mile supervolcano that erupted with explosive violence 29.5 million years ago. It is Oregon's largest, and very extinct, volcano.

stratosphere where it usually drifts far from the volcano before landing. They dispatch clouds of hot ash away from the vent at almost supersonic speeds to coagulate and cool elsewhere. Some of their ejecta goes straight up, and then comes straight back down to refill the great cavity left by the explosion. They leave little obvious evidence at the crime scene. But they are monsters when they erupt. Think supervolcano. Think 40-mile-high columns of ash. Think Yellowstone.

The largest volcano in the Northwest, 25 miles wide and 20 miles long, is an Oligocene caldera that erupted 29.5 million years ago. Known as the Crooked River caldera, its crater extends north to south from Oregon's Smith Rock State Park to Powell Buttes, and east to west from Prineville's Ochoco Reservoir almost to Redmond. Smith Rock is part of the towering north wall of the volcano. It is a place that dwarfs the human scale. The cataclysmic eruption produced the tuff of Smith Rock 29.56 million years ago. Just north of Prineville, Barnes Butte—a resurgent dome that filled the vacated caldera—also clocks in at 29.56 million years. Gray and Grizzly Buttes, slightly younger at 28.8 million years, are rhyolite domes erupted around the caldera's periphery after the eruption. The Crooked River caldera produced huge amounts of ash, now ensconced in the Painted Hills, including the well-documented Picture Gorge Ignimbrite, perched atop Carroll Rim, which is the same age and same composition as Smith Rock.

More Calderas
There are other calderas in central Oregon as well, including parts of the Ochocos, and adjacent areas, stretching from Prineville to Redmond to Warm Springs and Maupin. Steins Pillar, a 500-foot isolated pillar of rhyolite tuffs along Mill Creek about 15 miles east of Prineville, memorializes the explosive eruption that formed the older, 10-mile-diameter Wildcat Mountain caldera (39.4 Ma).

About 70 miles to the northeast, the Tower Mountain Volcanic Field covers about 200 square miles. Tower Mountain marks the remnants of a caldera that began as a good-natured, quiet vent, producing docile eruptions of fluid basalt, 30 million years ago. However, its Mr. Hyde personality slowly appeared about 29 million years ago as it vented increasingly explosive eruptions and voluminous mudflows. Between 29 and 28 million years ago (contemporary with the Crooked River caldera), explosive eruptions bored a 7-mile-wide caldera and sent pyroclastic flows racing across the landscape.

Other tempestuous eruptions occurred from as-yet-unnamed and unmapped vents near Ashwood and in the Mutton Mountains on Oregon's Warm Springs Reservation. These volcanoes are the likely sources of the gold, silver, and mercury deposits found at the Queen of the West, Horse Heaven, and other mines that pepper the Ochocos. Their gassy, vesicle-rich rhyolites host thundereggs and other rock-hound delights.

Why did these huge eruptions occur near the center of Oregon during the Oligocene? Hypotheses include links to a nascent Yellowstone hotspot lurking beneath newly thickened and accreted crust, or offspring of hot mantle or mantle-derived magmas forced to the north by the odd geometry of the dangling Farallon plate. The subject merits further investigation.

Steins Pillar is a remnant of the 40-million-year-old Wildcat Mountain caldera, exposed about 30 miles northeast of Prineville, Oregon.

The Cascades Begin: 38 to 25 Million Years
The Cascades of Washington, Oregon, and northern California began as a low range of explosive, ashy peaks in the Late Eocene. This early (38 to 12 Ma) volcanic range is known as the Western Cascades because most of its volcanoes lay to the west of the modern High Cascades peaks. Today, the Western Cascades volcanoes are long extinct, and most of their relics are buried deep beneath the modern Cascades.

Pilot Rock, south of Ashland, is among the oldest of the Western Cascades (and certainly the oldest Cascades-related volcanics in Oregon). Part of the Colestin Formation, Pilot Rock is a stubby vent/lava flow system about 38 million years in age. Another volcano of similar age is exposed at Tipsoo Lake, just east of Mount Rainier. The gray rocks exposed along the lakeside hiking trail reveal andesites erupted along the Pacific shoreline about 39 to 37 million years ago, as the Eocene closed and the Oligocene dawned.

The Oligocene also crops out on the western, southern, and eastern base of Mount Rainer as the Ohanapecosh Formation—a 2.5-mile-thick accumulation of gray andesites, stout mudflows, and thick layers of pyroclastic flows. On Rainier's west, these older, altered andesites and mudflows are exposed along the upper Puyallup River. Ancient vents and ignimbrites of this Late Eocene to Oligocene volcanic system include Cowlitz Chimneys, the Sarvent Glaciers area, and Mount Wow—all in Mount Rainier National Park. At Mount Wow, the stack of Oligocene volcanic rocks is more than a mile thick. Gray andesites peer from mossy outcrops on the southeast flank of Mount Rainier, and reappear along the Columbia River Gorge east of Cascades Locks and Bonneville. In Oregon, the Ohanapecosh makes a single, encore appearance—a low outcrop of undistinguished gray andesite at the trailhead to Latourell Falls—generally ignored by tourists intent upon visiting the glistening waters just down the trail.

Younger Oligocene volcanics occur east of Mount Rainier. The thick volcanic edifice of Fifes Peak is perhaps the most majestic. In its glory days, 27 to 25 million years ago, Fifes Peak was a huge composite volcano that covered an area larger than Mount Rainer. Its violent eruptions produced a Crater Lake–like caldera about 2 miles in diameter. Today, only the spire remains, a sliver of the north rampart of the once-dominant volcano.

Mount Wow (right, with Mount Rainier in the distance) looms over Tipsoo Lake. The peak is a remnant of an early Oligocene volcano that was among the first volcanoes in the Washington Cascades. It is considered part of the Ohanapecosh Formation.

Fifes Peak is the remnant of a large and very explosive Cascade volcano that erupted about 27 million years ago about 35 miles east of Mount Rainier.

Skinner Butte is a climber's mecca on the outskirts of downtown Eugene, Oregon. Although it looks like a lava flow, it is actually an invasive sill that burrowed its way through softer sediments, and then was exposed by uplift and erosion.

In Oregon, relics of Western Cascades volcanoes appear in the central Cascades in the canyons of the North and South Santiam River, at Bohemia, and above the McKenzie River's canyon. Tidbits Mountain once coughed ash east to the Painted Hills and John Day Basin where it fell heavily upon rhino-like brontotheres, scavenging entelodonts, and other residents of the east side's Oligocene temperate landscape. Two prominent buttes—remnants of Oligocene coastal volcanism—are landmarks in Eugene. The smaller, Skinner Butte, on the north fringe of downtown, sports prominent columnar jointing on its quarried west face, and helps define the city's downtown landscape. It is a favorite of neophyte rock climbers. The larger, Spencer Butte, rises to an elevation of 2,054 feet about 2 miles south of town, boasting a park, with an extensive system of trails that roam its flanks and climb to the summit. The butte evidently was called Cham-o-tee (rattlesnake) by the native Kalapuya tribe.

Both buttes were once considered basaltic intrusions. But geophysical surveys of their gravity and magnetic signatures indicate that neither have roots or dikes of dense igneous rocks that continue to great depths beneath them—as one would expect if they were actually true intrusive bodies. These buttes are surrounded by, and appear to intrude, shallow marine sedimentary rocks of similar age (40 to 35 million years ago, this was just offshore of the beach). It appears likely that the rootless intrusions that form Skinner and Spencer Buttes are invasive lava flows that sank into soft sediments on or off the shore and ponded beneath the surface, creating large masses of basalt and andesite that have since been exposed by erosion of the surrounding soft sediment.

The sea waves broke at the foot of the Oligocene Cascades. Evidence in Washington includes marine sediments in part of the Ohanapecosh Formation. In Oregon, there are Oligocene sea stacks on the east slopes of the Willamette Valley near Mount Angel, unveiled by Abiqua Creek's erosion. The fossils of barnacles still cling to wave-polished pinnacles of basalt, some 30 million years after a turbulent ocean sculpted these stumpy sea stacks along the beaches of an Oligocene sea.

At 35 million years ago, seafloor off Oregon and Washington subducted beneath North America more obliquely than today's head-on, due-east direction. The skewed convergence was slower. It also generated more faults, friction, and melting of near-surface crust. Consequently, as the melted rocks rose, they erupted—or in some places, solidified, stillborn, underground. Many of Oregon's coastal promontories, including Tillamook Head, Cascade Head, and the basalt at Yachats, as well as small, subtle intrusives of syenite at Table Mountain and Blodgett Peak in the Oregon Coast Range and the gabbro sill at Mary's Peak, are legacies of Oligocene volcanic and magmatic activity.

The Oligocene Coast, the Coast Range, and Marine Life

The modern Coast Range—a relict of the accreted Siletzia terrane—would be elevated slowly during the Oligocene, and as lands rose, the Willamette Valley would become less a bay, and more the valley landscape of today.

This shore, like ecosystems worldwide, was a cooler place. The ornate warm-water conchs of the Eocene were gone. Instead, utilitarian species, attuned to cooler waters and no-nonsense living, began to move in. The diversity that characterized Eocene marine communities vanished. Shellfish found in the Pittsburgh Bluff Formation in Oregon's northern Coast Range included animals like *Opalia*, adapted to

A clam shell accompanies a small shark's tooth (Sand Shark, *Odontaspis*?) in the Pittsburgh Bluff Formation in the northern Oregon Coast Range.

warm waters, as well as *Taranis*, a small, simple, thick-shelled gastropod found in deeper and cooler environments. Mollusks we can find today appeared by the end of the Oligocene, including razor clams (*Solen*) and the elongate shells of *Dentalium*, used as beads, decoration, and currency by coastal tribes.

Peculiar, isolated pods of soft limestones found in the Coast Range indicate that there were methane seeps offshore. Today, and presumably then, cold water methane seeps occur where gas is generated by the decay of organic material or by biological processes in sediments. As the methane (CH_4) reaches the top of the sediments, it reacts with oxygenated, calcium-rich seawater, producing limestone. Sulfur is frequently associated with the seeps, and helps produce an environment that supports dense biologic communities, including bacteria and algal mats that oxidize sulfur, and clams and tube worms that glean food from bacteria and algae. The limestones produced

at these seeps also provide a solid foundation for biological communities to grow. At modern seeps in the Gulf of Mexico, mounds of carbonate tower 1,000 feet above the seafloor. At the brink of the modern Cascadia Trench, methane seeps produce reef-like mounds of carbonate that support flourishing biological communities.

Limestone lenses near Mist, Oregon, seem an Oligocene equivalent. They are interbedded with the fine sandstones of the Keasey Formation. Two species of crinoids (*Isocrinus oregonensis* and *I. nehalemensis* are found sprouting from the limestone—or more often, occurring as fossils within the carbonates. These odd echinoderms, also known as "sea lilies," are a sort of sea anemone on a long stalk. Though rare today and uncommon during the past 66 million years, crinoids were bounteous during the Paleozoic and Mesozoic. Their occurrence in these limestones suggests an odd chemistry and a rare occurrence of solid carbonate hard-grounds rather than soft sand to which the crinoids' slender feet could cling.

Near Naselle, Washington, more fossilized methane seeps of Late Eocene to Oligocene age contain fossil sponges galore. In the Lincoln Creek Formation, exposed near Holcomb, Washington, along the Canyon River on the southern Olympic Peninsula, James L. Goedert and colleagues of the Natural History Museum of Los Angeles found corals. The corals, Goedert notes, were likely attracted to the methane seeps by the heady combination of abundant food and a hard, stable carbonate substrate they could hold on to.

There is little record of fish in the Northwest's Oligocene marine sediments, although sharks—including *Hexanchus*, a blunt-nosed, six-gilled, deep-water shark, and *Notorhynchus*, a large seven-gilled shark and shallow-water specialist—left their teeth and other remains here for paleontologists to find. The Oligocene is also when megaladon, a monster shark that grew up to 60 feet long appeared, but no known traces of this long-extinct mega-shark have been found in Northwest waters.

Whales are the real stars of the Northwest's Oligocene marine fauna. Distant cousins of cows and camels, they appeared in the Paleocene and Eocene, including huge versions like *Basilosaurus*—40 to 75 feet in length, with vestigial hind limbs, found on the US Gulf Coast and other localities globally, but not known in the Northwest. In the Oligocene, as global oceans cooled, the diversity of plankton species (not overall planktonic biomass) decreased from more than a thousand in the Late Eocene to a low of only about 500 species (30 Ma) before rebounding to 1,500 to 2,000 species known in the Miocene (15 Ma). Whales followed this trend. In the Eocene, fifteen genera of whales are known. In the Miocene, more than 130 genera flourished. But in the Oligocene, whales hit a low point of diversity, with only five genera known globally.

The best-known Oligocene fossil whales found in Oregon come from a locality near Yaquina Head (collected by Douglas Emlong),

Clams, gastropods, and crinoids occupied the fertile waters around a methane seep on the Oligocene seafloor, now exposed on the southern slopes of Radar Ridge in the southernmost Willapa Hills/Coast Range.

and from shallow-water sediments known as the Butte Creek Beds about 20 miles east of Salem. Bill Orr, who discovered the animal in the Butte Creek Beds, notes that both whales are aetiocetids, or baleen whales. They have blowholes halfway up their snouts and are considered transitional between ancient whales and modern whales, which have blowholes near the top of their head. Wanting to hedge their bets on what might be available on the local, plankton-poor menu, aetiocetids had a mouth full of teeth as well as baleen, as is characteristic of the modern blue whale today. Paleontologists think that the Oligocene whale's diet included not only plankton, but also tastier fish and crustaceans.

The fossils of whales from Washington's northwest Olympic Peninsula are even more intriguing. In the Makah and Pysht Formations, the remains of eight Oligocene whales were found in deep-water Oligocene sediments. This mass grave represents a whale fall—a place where whales either go to die, or where a group met their end 30 million years ago. The fossilized worms and mollusks found growing on the whale bones are the same ones found at deep sea vents and black smokers. They are chemo-symbiotic; they produce energy from sulfides. Paleontologist James Goedert of the Natural History Museum of Los Angeles and colleagues consider this evidence of not only whale falls from long, long ago, but also of bounteous sulfide production as the whale carcasses decayed—a process similar to whale falls today. Other bones of smaller whales, less than 12 feet in length, from the same formation preserve the teeth marks of scavenging sharks, as well as the fossilized remains of a large marine worm, *Osedax*, that has evolved with whales and consumes whale bones on the seafloor.

Changing Terrestrial Forests:
The Legacy of a Cooling Climate

In the cooling days of the Oligocene, the Northwest's lush jungle greenness faded to a more withered and seasonal savannah tan. The chill of Antarctic snows had a global reach. In Oregon and Washington there were seasons. We know this by the seasonal growth rings laid on by trees with increasing enthusiasm as the chill polar climates tightened their grip on the planet.

The Northwest's transition from tropical to temperate (39 and 32 Ma) is recorded in two places. Sediments and tuffs of the Western Cascades near Eugene and Goshen, Oregon, preserved the leaves of a changing forest, while soils of the Painted Hills near Mitchell, Oregon, recorded temperatures and precipitation in isotopes and clays. Both indicate a slow cooling trend, from about 72 degrees F as mean annual temperature in the Late Eocene to a mean annual temperature of about 55 degrees F with pronounced seasonality by 32 million years ago.

Leaf morphology is a generally reliable climatic indicator. Thick, smooth-edged leaves with pointed tips indicate a tropical environment; thinner leaves with more serrated edges and rounded tips characterize temperate forests. At the Eocene-Oligocene boundary, 34 million years ago, the coastal forest near Goshen consisted of large, thick, smooth-edged leaves with pointed tips that quickly shed rain: liquidambar (sweet gum), magnolia, katsura, and fig. Leaf morphology of the Goshen and similar localities, analyzed using the Climate Leaf Analysis Multivariate Program (CLAMP), suggests an average winter temperature of about 45 degrees F, and average summers of 76 degrees F. Today, average temperatures at nearby Eugene are 53 degrees F as the average annual temperature, 44 degrees F for winter, and 63 degrees F for summer.

Fossil leaves found near Lowell, Oregon, in the Coburg Hills, and south of Eugene along the Willamette River, demonstrate that by 30 million years ago, the climate had cooled even more. These woodlands are dominated by oak and metasequoia. Other trees included

The oldest portions of the Painted Hills include Red Hill, where red, tropical laterite soils that developed about 40 million years ago, in the warm, moist climate of the Late Eocene, are exposed.

The ancient soils exposed in the Painted Hills indicate that as time passed, the climate became cooler and drier. Red and tawny-yellows soils of moist, warm climates near the bottom of the sequence give way to progressively lighter-colored soils of cooler, drier times, upward in the sequence.

ancestral maple, pine, spruce, and walnut. According to leaf morphologies, the average annual temperature was 56 degrees F, with winter averages a chillier 42 degrees F, and summer averages at 68 degree F.

The Painted Hills near Mitchell, Oregon, provide a mineralogical and isotopic record of cooling and drying that corroborates the story in forest leaves. Soils of the Eocene are deep red tropical laterites, virtually identical to modern soils of Costa Rica or southern Mexico. Their clays and oxygen isotopes record warm temperatures and high rainfall. At 33 million years, the very beginning of the Oligocene, soils are still wet and climate is barely seasonal, with precipitation of about 50 to 60 inches per year, according to Greg Retallack and colleagues. Vegetation in the Painted Hills was dominated by walnut and laurel with metasequoia, grasses, and flowers in wetlands, and alders on drier upland slopes. By 31 million years ago, the ancient fossilized soils are no longer red and highly oxidized. They are brown andosols, the scions of temperate climates; the sort of soil found in mid-altitude grassy forests of Italy or Spain today, where the summer rain is a critical ecological force. Gone are the indicators of the tropics. Annual precipitation, which falls largely during the summer, drops to 25 to 35 inches. Bring on the oaks.

Grasslands and Climates

Grass—including grains such as wheat, barley, and corn, as well as rice, sugarcane, and bamboo—is essential for the welfare of humans.

Metasequoia, Oregon's State Fossil, prospered in the temperate climate of the Northwest's Oligocene.

Although this unassuming plant probably first appeared in the Late Cretaceous, the Oligocene was its debutant ball.

As climates cooled, deciduous became the de rigueur type of tree and forests became less dense. Metasequoias, alder, and beech flourished along creeks and wetlands. In open spaces, where there was room for smaller plants, grass seized the day. By 30 million years ago, bunchgrasses were well established. By 25 million years ago, sod-forming grasses were extant. In the John Day Basin, notes Greg Retallack, fossilized roots and root traces in soils provide evidence of bunch-grass. Herbivores had adapted to munching on grass, too. Horses and rhinos with high-crowned teeth designed for eating grasses appear and prosper in the early Middle Oligocene, about 30 million years ago.

The advent and spread of grass likely pushed planetary climates into even cooler territory for a few reasons. First, grass has a higher albedo than the dark, rough visage of a forest, reflecting almost 50 percent more light and heat back into space. Second, with drier and cooler climates, a grass that was more efficient at using and sequestering carbon developed—the C4 grasses. Essentially, these plants retain atmospheric carbon dioxide longer than more "normal" C3 grasses (the grass in your lawn) allowing them to process and produce more oxygen and photosynthetic energy, produce more sugars, and sequester more carbon. In the Oligocene, 26 to 62 percent of the grass pollen was from C4 grasses, according to Michael Urban at the University of Illinois. Plants helped reduce atmospheric CO_2 and lower global temperatures.

Horses, rhinos, elephants, and antelope also figure prominently into the rise of grasses. This Oligocene fauna developed a hankering for a new favorite food. Greg Retallack notes that for grass to spread, it needed a mechanism to control trees and also disperse its seeds. Fire comes to mind, but there is little evidence of forest fires during the time that grass rose to prominence. Seasonal dryness that would stress trees but not grass also comes to mind. But again, the mid-Oligocene, when grass increased dramatically, was actually a wetter time than the Early Oligocene—a pre-grass world. Retallack's preferred mechanism for the ascendancy of grass is the coevolution of grasses and grazers. In his words:

Grasses . . . are better adapted than other plants at withstanding the grazing pressure of large herds of ungulates. Horses and antelope, on the other hand, are uniquely suited by virtue of their high-crowned teeth, hooves and elongate limbs to life on the open plains. Large herbivores such as rhinos and elephants are particularly destructive of trees, stripping their bark and toppling their trunks. By this view, grassland sod evolved as a group of adaptations in roots and shoots to withstand increasingly effective trampling and grazing by mammals. Against a near-chaotic background of mountain uplift, sea level change and paleoclimatic oscillation, sod-grassland ecosystems appeared and stayed in semiarid regions of Oregon.

Animals Find Their Legs

Just north of downtown Fossil, Oregon, fragile, thinly bedded shales provide a glimpse of a 30-million-year-old lake. Here, oaks, maples, alders, and sycamores shaded the lakeshore, along with pines and metasequoia. Salamanders scuttled under leaves. Thick-skinned, bad-tempered, and foul-breathed entelodonts grudgingly shared the watering spot with rhinos and herds of oreodonts.

The temperate Oligocene fauna was substantially different than the animals that inhabited the semitropical Eocene forests. The animals of the coming kingdom of grasses were leggier, swifter, and more svelte. This was a more open landscape where pursuit counted more than pounce at mealtime, and the fauna adapted to the new hazards and opportunities.

Horses came into their own in the Oligocene. Most bet heavily on grass, and less on soft munchy forest leaves. *Anchitherium* was a three-toed forest-based leaf-browser that stood about 10 hands (40 inches) tall at the withers. (A "hand" is a horseman's unit of measure that probably originated in ancient Egypt. In Britain the measure was standardized by Henry VIII as 4 inches—the approximate width of the average man's hand.) *Anchitherium* represents a dead end on the branching paths of evolution. *Mesohippus*, with three toes, but one longer dominant middle toe, was the first to develop teeth that could chew grasses. It stood about 6 hands (24 inches)—about the size of a German Shepherd—and disappeared before the end of the Oligocene. *Miohippus*, a larger model that appeared about 36 million years ago, coexisted with its little cousin much like white-tailed and mule deer share today's landscapes. *Miohippus* was a true grassland native, with ankle joints better adapted to running on hard grassland surfaces, and high-crowned, hard-enameled, grass-cutting hypsodont teeth. Although its smaller, browsing cousin, *Mesohippus*, would fade into obscurity by 25 million years ago, *Miohippus* and its ensuing genera galloped into the future.

Many other animals were eager to try out the new grassy diet. There were at least five kinds of camel—ranging from goat- to llama-sized. *Paratylopus*, an abundant cameloid, ranged across the Midwest as well as the John Day Basin. Its goat-sized skeletal structure suggests that it weighed about 100 pounds. *Gentilicamelus* was only slightly bigger. *Oxydactylus,* a 400-pound hoofed camel, was the most striking of the Northwest's Oligocene

camels. Standing 5 feet at the shoulder, these animals had long legs and necks, which, like modern giraffes, allowed them to browse trees.

The Oligocene saw the advent (and demise) of sheep-like oreodonts. These animals, slightly larger than average modern sheep, had short legs, five-toed feet, and teeth made for eating grass. Like sheep, their teeth suggest that they were ruminants, though with long canine teeth some oreodonts may have been omnivores. They gathered in large groups, roaming North America like herds of tiny buffalo. At least sixteen different oreodont genera occupied the John Day Basin during the Oligocene—a diversity greater than the genera of antelope and deer in all of modern North America. They were so abundant in the John Day Basin that discovery of their fossils today stirs zero excitement among John Day Fossil Beds National Monument paleontologists.

Entelodonts, aptly named "terminator pigs," were buffalo-sized, gnarled, and toothy wild boars with bad tempers. Their heads sported tusks for fighting. Entelodont skulls often display chips and healed scars made by the teeth of other entelodonts during battles. The pattern of wear on entelodonts' teeth and their odd jaw motion suggest a bizarre diet. Their front teeth were evidently used mostly for crushing very hard material. Their orbital jaw mechanics and strong jaws suggest that they might have consumed a high percentage of bone and carrion. Entelodonts, it seems, were mostly scavengers. Oddly, entelodonts abruptly disappeared from the North American record 25.8 million years ago, along

with cat-like nimravids. They both reappeared just as abruptly 18 million years ago. No one knows why. (Perhaps the entelodonts went to a much-needed therapy session, and the nimravids simply took a nap.)

In a world where herbivores proliferated, meat eating grew popular. Carnivores diversified to take advantage of increasingly diverse menu items. Some early carnivores were similar in appearance and ecological function, but genetically unrelated, to modern animals. These included nimravids—cat-like animals that had partly retractable claws and lacked some auditory, cranial, and other characteristics of modern cats. In the Oligocene Northwest, at least five different genera of nimravids prowled the oak savannahs: *Nimravus*, a fossil identified by Cope in 1879, was a lithe animal about 4 feet long that weighed about 65 pounds; *Hoplophoneus*, a bobcat-sized creature that sported daunting 6-inch-long saber teeth; and a more dainty *Dinictis*, whose fangs were a diminutive 2 inches; all prowled what is now central Oregon. The much larger, leopard-sized *Eusmilus*, with saber teeth of 6 to 8 inches in length, also ranged here. *Eusmilus* had short legs, a body up to 8 feet in length, and weighed about 150 pounds. In the John Day Fossil Beds, a *Nimravus* skull was found with a hole in it that matched the size of the average *Eusmilus* saber tooth. The story has a happy ending: the nimravid skull had healed the wound, with substantial bone growth sealing the hole, so the smaller animal evidently survived the painful encounter.

Bears migrated across a land bridge from Asia during a time of lowered sea level in the Early Oligocene. These animals are more properly considered "bear-dogs," part of the subfamily of Amphicynodontinae. They include *Parictis,* the oldest known bear, inhabiting the Northwest and western North America from 38 to about 33 million years ago. It would be hard for us to recognize *Parictis* as a bear. It was a slender animal with a head only about 3 inches long. But we would recognize the bear-like nature of the approximately 29-million-year-old *Allocyon,* whose heavy, blunt, and toothsome jaw and skull were found in 1920 by Charles Merriam near Logan Butte, south of John Day, Oregon. This animal had a head about nine inches long, and is characterized as a medium-sized predator by the University of California, Berkeley, paleontology department.

There were dogs as well. At least nine different canids (true dogs) roamed the Northwest during the Oligocene. The most widespread and abundant was *Mesocyon,* found in the Turtle Cove Formation of the John Day Fossil Beds. Weighing about 15 pounds, with the graceful lines of a fox, *Mesocyon* was the first canid "hyper-carnivore"—an animal specialized in eating meat. Its unfused wristbones may have allowed it to climb trees. It shared the landscape with the Chihuahua of the day—a tiny, 3-pound canid known as *Hesperocyon.* On the other end of the scale, muscular *Osbornodon* stood about 20 inches tall and weighed about 50 pounds. It had

short but powerful jaws and heavy molars, and belonged to an extinct (thank goodness) group of canids called Borophaginae—or "bone crushers." A much smaller "bone-crushing" canid, *Cormocyn,* shared the Northwest's grassy savannah. *Cormocyn* would have been equivalent in size (and possibly attitude) to a Jack Russell Terrier, and weighed an estimated 6 to 7 pounds.

And, of course, there was the charismatic mini-fauna, which included the first rabbits (less than 4-inch-long *Palaeolagus* and *Archeolagus*), squirrels, gophers, field mice, and beavers. The lemur-like arboreal animal *Ekgmowechashala* (from the Oglala Lakota word meaning "little fox man" or "little cat man") was first discovered on Wazí Aháŋhaŋ Oyáŋke (the Pine Ridge Reservation), and confirmed as a Pacific Northwest resident from a single tooth found in the John Day Fossil Beds. The animal was about 1 foot long, weighed perhaps 5 pounds, and thrived on soft fruits that included early apples, pears, and apricots.

We will never have a complete catalogue of the animals and plants of the Oligocene. Or any other time. The fossil record is, by its nature, incomplete. Our understanding depends entirely on chance discovery by a person who is knowledgeable enough to recognize the significance of a fossil. But our limited view of the Oligocene, 34 to 25 million years ago, provides a reasonable idea of a Northwest landscape at once familiar and strange.

The distinctive color of the Oligocene Turtle Cove Formation is produced by the mineral celadonite, a blue-green, low-temperature clay. These beds were deposited in wet conditions, and preserve an astounding array of Oligocene animals, including entelodonts, oreodonts, brontotheres, nimravids, and, yes, turtles!

During the Miocene, much of the Northwest was inundated by basalt lava flows. Lookout Mountain, south of Baker City, Oregon, was a site of some of the eruptions.

Global Temperatures

Millions of years

2500 PROTEROZOIC 541 PALEOZOIC 252 MESOZOIC 66.0 CENOZOIC

Camb | Ordo | Silur | Dev | Miss | Penn | Perm | Trias | Juras | Creta | Paleo | Eoce | Olig | Mioc | Plioc | Pleist | Holo | Anth

CHAPTER 8   The Great Lava Flows   *Basalts Flood the Landscape*

The Miocene spans the time from 23.03 million years ago to just 5.3 million years ago. Its legacy of vast basalt flows forged much of the Northwest's modern landscape. Travel through the Columbia River Gorge, and you are cradled in the Miocene's dark lavas. Explore Steens Mountain and you stand atop its eruptive fury. Tour the Oregon coast, and witness its sizzling connections with the sea. The Cascades grew, the Basin and Range began to open, grasslands dominated, and horses got bigger. But most importantly, the Columbia River basalts (CRB) spread dark, fluid lava as much as three miles thick across most of the Northwest.

Early Miocene: The Cascades Grow (23 to 16 Ma)

In Washington, only the roots of Miocene Cascade volcanoes remain. These include parts of the large Snoqualmie batholith, which is exposed over 270 square miles in the mountains east of Seattle. This now-solidified magma chamber was active 28 to 17 million years ago, operating mostly in the Miocene. It fed volcanoes which have long-since eroded away, although nearby rhyolites dated at 20 million years suggest that not every vestige of volcanic rock has disappeared. Cloudy Pass pluton north of Mount Rainier is about 22 million years in age. The Tatoosh pluton, just east of Rainier, spans the time from 26 to 14 million years of age. In the early Miocene, 20 and 22 million years ago, the Tatoosh magmas reached the surface, erupting as thick ash flows.

In Oregon, the early Cascades are represented by several formations. The Little Butte volcanics' demure volcanoes produced relatively small volumes of tuff, ash, and andesitic lavas.

Near Bohemia, vents produced andesites and more mafic volcanism, while their near-surface magma chambers deposited gold, tin, lead, and silver mined commercially from about 1870 to 1942. Closer to the Columbia Gorge, volcanoes of the Eagle Creek Formation disgorged andesites, dacites, and the voluminous mudflows of the Eagle Creek Formation exposed in the crumbling cliffs of Table Mountain and the vast Bonneville (Bridge of the Gods) Slide above Stevenson, Washington.

The Early Miocene lahars and ash beds captured a record of Early Miocene forests. It was a landscape with an increasingly evident rain shadow. Oregon's pioneering paleobotanist, Ralph W. Chaney, noted the disparity between eastern Oregon's dry-side vegetation and the wetter, west-side forests in his 1918 treatise, *The Ecological Significance of the Eagle Creek Flora*. The Miocene was the first time that a distinctly dry climate appeared on the east side of the Cascades, including the appearance of more desert-loving plants like ancestral sage.

Basalts Inundate the Landscape (16.9 to 5.5 Ma)

If you have ever strolled across the rippled black slopes of Kilauea, or explored the once-molten surface of Craters of the Moon, then you have an inkling of what Pasco,

Washington, might have looked like 15 million years ago. Long before Kennewick Man, Columbian mammoths, or Ice Age Floods, the Columbia Basin was a flat, desolate, dark expanse from horizon to horizon. There was no Rattlesnake Mountain. No Badger Mountain. No Horse Heaven Ridge, Umtanum Ridge, or Vancycle Ridge. No Blue Mountains, and no Wallowas to rescue the landscape from its somber monotony. No Wallula Gap to usher in the winds.

Depending upon where your time-travel capsule lands on this Miocene landscape, you could find some hardy grasses and resilient ferns emerging from the barren ground. Perhaps an oak tree or maple or metasequoia sprouting bravely from ebony-colored rocks. But from horizon to horizon, everywhere you can see, is a landscape of basalt.

In the time since these basalt flows appeared, the Northwest has been faulted and uplifted, folded and torn. Rivers sliced new channels through the basalt flows. Ice Age

winds strew them with dunes. Ice Age Floods sculpted and eroded a channeled topography into their layer-cake landscape. Despite nature's efforts to cover them up or scour them away, the vast lava flows of the Columbia River Basalt Group remain the seminal and ubiquitous underpinnings of our landscape.

They are a type of basalt formation known as flood basalts. Rare in Earth's history, flood basalts spring directly from the Earth's mantle. Often their source is a mantle plume or hotspot. They erupt in mind-boggling volume,

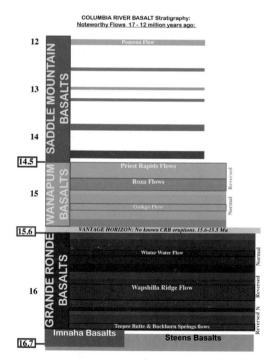

Columbia River basalts include voluminous basalts erupted from vents in southeastern Oregon on and near Steens Mountain.

The Columbia River flood basalts inundated much of the Pacific Northwest during the Miocene. In the steep, open landscapes of eastern Washington and Oregon, including this view across the Imnaha Canyon and Hells Canyon, their extent is quite evident.

"flooding" the landscape with basalt. Examples include the Siberian Traps (252 Ma) that coincided with the greatest extinction on the planet; the 201-million-year-old basalts of the Central Atlantic Magmatic Province (CAMP) implicated in the end-Triassic extinction; the 180-million-year-old Parana Basalt of South America and related Karoo Basalt of Africa—both generated as the southern Atlantic Ocean began opening; and the Deccan Traps in western India that coincided with the extinction of dinosaurs 65.9 million years ago.

Our backyard flood basalts are known as the Columbia River Basalt Group, or, more informally, the Columbia River basalts, or CRB, Steve Reidel, of the Pacific Northwest National Laboratory, notes that the CRB are the youngest, smallest, and best preserved of all known flood basalt provinces on the planet.

A Brief Chronology of the Columbia River Basalts

Beginning 16.9 million years ago, great volumes of this hot, fluid lava erupted from long, linear vents, mostly in eastern Oregon and eastern Washington. Lava inundated 80,000 square miles of Oregon, Washington, western Idaho, and northern Nevada. More than 98 percent of the 350 great CRB lava flows erupted between 16.7 and 15.2 million years ago, according to studies by T. L. Barry, Stephen Self, and others. Most of this spurted out of the ground in only 420,000 years, between 15.99 and 15.57 million years

ago. Major lava flows appeared, on average, every 4,000 years. Sulfur and ash from the largest eruptions may have generated brief "nuclear winters." The carbon dioxide released also produced a global spike in temperatures known as the "Middle Miocene Thermal Optimum."

In all, about 50,000 cubic miles of basalt inundated the flat-lying landscape of eastern Oregon, eastern Washington, western Idaho, and part of northern Nevada. This is enough basalt to build an interstate highway to the moon—a strip of basalt 238,000 miles long, 100 feet wide, and 10 feet thick (or a hiking path 10 feet wide and 100 feet thick).

Source and Generation of Columbia River Basalts

What generated the great volumes of lava that make up the CRB? No single mechanism is universally accepted. Three technically sophisticated stories vie for the best explanation: a rent in the down-going oceanic plate; the Wallowas jettisoning their deep, crustal root; or a link to the Yellowstone hotspot.

The most recent idea is that a rip in a huge chunk of oceanic crust that had been subducted deep beneath today's Oregon-Nevada state line allowed mantle-derived magmas to rise toward the surface in great volumes. The idea was proposed by Lijun Liu and Dave Stegman of Scripps Institution of Oceanography.

Their story is that about 17 million years ago, and almost 40 miles beneath McDermitt, Nevada, the Farallon plate was having a bad

day. This oceanic crust had been pulled down into the subduction zone beneath the western United States. But a hot plume of rising basalt began to weaken the slab, producing an opening that the magma could begin to move through. As the hot basalt continued to work its way through the 3-mile-thick Farallon slab, this chunk of cool oceanic crust became more flexible and more eroded. Eventually, the warming slab of oceanic crust began to buckle and then tear apart.

Thin fingers of basaltic lava pierced the weakened slab and moved toward the surface. About 2 to 3 miles beneath the ground, heat and upward pressure began to melt the lower crust, producing a magma chamber charged with water and other volatiles. As the melt became more fluid, the top of the magma chamber collapsed into the maelstrom of magma below. The ensuing explosive eruption of ash and ignimbrites created the 15-mile-diameter McDermitt caldera 17 million years ago and heralded the onset of massive eruptions of basalt across eastern Oregon and eastern Washington.

The initial rupture of the Farallon plate, and eruption of basalts at McDermitt caldera and Steens Mountain, may have been foreshadowed even earlier by explosive eruptions of tuff and rhyolite from flat, caldera-like vents. These light-colored rocks are known as the Alvord Creek Beds (21.5 Ma) and Pike Creek volcanics (21 Ma). These light-colored rocks form the base of Steens Mountain along its eastern and southern faces. They represent

Columbia River basalt eruptions began in southeastern Oregon where Steens Mountain is today. The light-colored outcrops in the foreground are part of the Steens Volcanics—rhyolites that were the precursors to the vast basalt eruptions.

Yellowstone plume by the deep and detached root of the Wallowa Mountains could account for the Columbia River basalts.

Lui and Stegman's 2012 explanation that a foundering Farallon plate generated the CRB is now the most widely accepted. It fits the chemistry, geography, and timing of eruptions better than the rest. It also supports the newly accepted understanding that the basalts erupted at Steens Mountain in southeast Oregon, 16.9 to 16.6 million years ago, were actually the first CRB lavas, rather than a separate, Yellowstone-related volcanic province.

The Far-Traveled Flows

Why were CRB volumes so huge? Work by Rebecca Lange of the University of Michigan suggests that the CRB magmas contained at least three to four times more steam and carbon dioxide than Hawaiian lavas. Such high concentrations of volatiles help melt more of the source area, and also help drive the process of eruptions, decreasing the density of the lava, helping it rise faster, and evacuating more of the magma chamber. The higher concentrations of carbon dioxide may have also helped in creating the mid-Miocene thermal optimum about 15.8 million years ago—a time, according to Wolfram Kuschner and colleagues at Utrecht University, when atmospheric CO_2 reached 500 ppm and global temperatures rose briefly.

And last but not least is the intriguing question: How did these lavas manage to

lower crust that was melted and then partly mixed with the ominous, rising plume of basaltic magma. As the Farallon slab continued to rupture, a growing fracture allowed more and more basaltic lava to reach the surface. Ultimately the Farallon slab itself melted, adding even more basaltic material to the ascending plumes. Then, 16.9 million years ago, the basalt reached the surface. And the Northwest's landscape changed forever.

Previous models for generation of the CRB involve mantle plumes, including Yellowstone.

In the 1980s, Alan Smith proposed that the CRB were the products of back-arc magmatism. The idea is that a down-going oceanic slab stirs up the mantle, producing volcanoes "behind" the active volcanoes—in this case, the Cascades. But no back-arc setting has ever produced large volumes of basalt. In the 1990s, Peter Hooper invoked the Yellowstone plume. The dates and generally the chemistry fit, although it was hard to explain why the bulk of CRB lavas erupted far north of the Yellowstone hotspot's track. In 2008, T. C. Hales suggested that deflection of the flow so far? We tend to think of a lava flow as a molten river. In videos of Hawaiian lavas, the incandescent orange fluids of pahoehoe sometimes ooze, and often race, through dark gutters of black stone. Or more solid lava (aa) advances deliberately in an unstoppable wall, glowing and bleeding like a special-effects monster as huge rocks tumble onto hapless innocents below. Neither seems sustainable over hundreds of miles.

Hawaiian flows actually do offer a mechanism for how the CRB flows moved. It's known as "inflated pahoehoe." As flowing lava cools, it develops a thin skin of glassy basalt on the top. As more and more lava flows under this skin, the flow inflates, like a balloon. The inflated portion of the flow may be tens of feet high. But ultimately, like a balloon at the end of a garden hose, the insulating, restraining skin bursts, usually at the front of the lobe. New lava surges forward, repeating the process again and again, across hundreds of miles. The high temperatures of all the recently solidified basalt also helps lavas retain heat, and acts as an insulating blanket for the rest of the flow to move across.

This process is observed at Kilauea and other active basalt vents. Its frozen equivalent is obvious in recent lava fields including Craters of the Moon in Idaho and Diamond Craters in Oregon. And it is apparent in many Columbia River basalts, where multiple chill zones, layers of columnar jointing, and preserved lobes record this history.

Inflated flows advance slowly—at rates of 0.5 to 7 mph—because much of their volume is used to thicken the flow as they advance. Volcanologist Stephen Self calculates that it would take weeks, or possibly years, for an inflated flow to move from eastern Oregon to the Pacific. But they did.

Columbia River Basalt Stratigraphy

The Middle Miocene topography of the Pacific Northwest was substantially different than it is today. To understand why CRB lavas were able to spread so far, we need to step back into our time machines and emerge into the Northwest as the first Steens eruptions begin. This is a gentle landscape—a place that more closely resembles Kansas than Oregon. There is no Coast Range, no Olympic Peninsula. The Willamette Valley is a tidal marshland west of a low, mostly extinct Western Cascade volcanic range. Streams thread their way from this old, subdued set of peaks directly to the Pacific. Some drainages flow from the east side through these derelict Western Cascade peaks directly to the ocean. The modern High Cascades are absent. They will not begin to erupt for another 12 million years. That means no Mount Hood, Jefferson, Rainier, Adams, or St. Helens. The Columbia River runs diagonally across the interior of Washington, then through the place where Mount Hood will rise in the future, and meets the Pacific perhaps 20 miles west of Salem, Oregon.

In southeast Oregon, there are no towering rims or arid basins. Instead, you find a grass-laden, oak-speckled plain. The area around Steens Mountain 16.7 million years ago was a low but extensive shield volcano, perhaps similar in appearance to Newberry Volcano today. In northeast Oregon, the most impressive topography may be Tower Mountain—a lone Oligocene volcano near Ukiah that might have offered a few thousand feet of relief. But elsewhere the landscape consists of undulating, gentle valleys and extensive stretches of flatness. There are no Elkhorns or Wallowas or Seven Devils. The great gouge that is Hells Canyon today, instead is a modest dale.

The Columbia River Basalts Eruptive Phases

The Columbia River basalts erupted in seven major pulses: Steens, Imnaha, Grande Ronde,

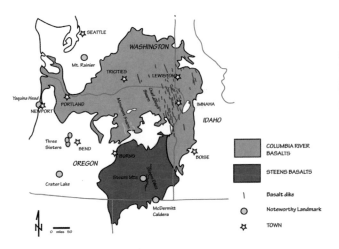

Columbia River Basalt stratigraphy (simplified).

Prineville, Picture Gorge, Wanapum, and Saddle Mountain. Each pulse consisted of tens to hundreds of major flows. Each has a distinctive geochemical signature and a slightly different genesis. Each is named for a location where it is well exposed.

In formal geologic terms, the Columbia River basalts are officially the Columbia River Basalt Group (CRBG). Each of the seven pulses is considered a separate, formal, geologic formation (in the stratigraphic sense that a formation is a contiguous series of rock strata with similar ages and compositions that were deposited in similar conditions by similar processes). Within the formations, "members" consist of individual or multiple flows with distinctive compositions. So, for example, the largest of the CRB flows is the Wapshilla Ridge flow that erupted from a long, linear vent now exposed near Maxwell Lake in the Oregon's Wallowa Mountains. Its formal geologic designation would be the Wapshilla Ridge Member of the Grande Ronde Formation, Columbia River Basalt Group.

Because this book is intended for readers who are not professional geologists, more informal terminology will be used throughout the discussion of CRB flows. However, understanding the overall order of things and the overall unity of the CRBG—the seven major pulses or episodes of eruptions (formations), each with multiple sets of flows (members)—is important.

Vent Systems

Most CRB flows erupted from long, narrow linear vents (fissures), often 10 feet wide or less, and 10 to 60 miles long. The eruptions were similar to the fire-fountaining fissure eruptions in modern Hawai'i—only much more formidable and longer lasting. The largest flows likely erupted from the same vent system for years to decades, spewing lava thousands of feet into the air. When the eruption ceased, basalt lava solidified in the vent, forming a dike. Today, some of these basaltic dikes form resistant "walls," or, especially in the Wallowa Mountains, dark streaks that contrast with white granitic rocks and light gray limestones.

Each phase of CRB eruptions originated from vents within a geographically defined source area. Geologists call these clusters of basaltic dikes "dike swarms." They represent places where rising plumes of magma forced their way to the surface, through cracks or fissures.

Three of the eruptive phases appeared in relatively small and well-defined areas. The Steens basalts erupted from what would eventually become the faulted, uplifted massif of Steens Mountain. The Prineville basalts erupted from fissures in east-central Oregon near Prineville Reservoir. These elongate vents are the Prineville basalt dike swarm. The Picture Gorge basalts, exquisitely exposed at Picture Gorge along the John Day River, actually erupted near the town of Monument, Oregon. This source area is termed the Monument dike swarm.

Two dikes of Wanapum basalt are exposed along the Grande Ronde River near the Oregon–Washington state line.

The most voluminous CRB phases—Imnaha, Grande Ronde, Wanapum, and Saddle Mountain basalts—share a large geographic vent area. Known as the Chief Joseph dike swarm, after *Hin-mah-too-yah-lat-kekt* (Chief Joseph) of the Nimi'ipuu (Nez Perce), it occupies an elliptical area of eastern Washington, eastern Oregon, and western Idaho. Bill Taubeneck of Oregon State University named the swarm and estimated that it contained about 21,000 dikes. More than 70 percent of the CRB volume originated from the Chief Joseph swarm.

The Steens Basalt

The Steens basalts are the oldest CRB lavas, and make up about 15 percent of all the CRB. They erupted between 16.9 and 16.6 million years ago, according to T. L. Barry of the University of Leicester. The most intensive activity occurred about 16.7 million years ago. It is likely that the entire thickness of basalt exposed on Steens Mountain erupted in 100,000 years or less, according to Victor Camp and colleagues of San Diego State University. The lava flows, and many dikes that fed the eruptions, are elegantly exposed as vertical stripes of dark rock in the precipitous east face of Steens Mountain.

Steens Basalt covers 20,400 square miles with a volume estimated as 7,600 cubic miles. That is more than enough basalt to fill a tunnel a mile wide and a mile high that stretches from Seattle to Miami and back. The lavas extend east to Lakeview, Abert Rim, and

beyond, as far as Table Rock in the Fort Rock Basin. They glower from Hart Mountain and Poker Jim Rim's dark cliffs, from Abert Rim's steep face. From Steens summit, the basalts continue beneath the floor of the Alvord Desert north and east at least 40 miles to New Princeton and Ventanor. To the south, they cover parts of Nevada, including Oregon Canyon and adjacent areas.

The faulting that uplifted today's Steens Mountain did not begin until several million years after eruptions of Steens Basalt ceased— or about 14 to 15 million years ago. The Alvord Fault has uplifted the massive mountain more than 4,000 feet above the Alvord Desert, and providing a cross section to decipher its history. The dissected volcano reveals not only its

Early Miocene, light-colored, more rhyolitic base, but also the network of narrow dikes and conduits that conveyed lava through the growing pile of basalts to erupt near the giant shield volcano's summit. They are dark stripes that slice upward through the mountain's steep, east-face layer-cake of basalt flows.

Some basalts on Steens Mountain, Abert Rim, and locations in between contain very large plagioclase feldspar crystals, up to 3 inches in length. They occupy up to half the volume of the rock, with their white or clear patches producing an odd, highly visible pattern. The distinctive texture has earned them the name "turkey-track" basalts. These rocks are the result of lava that lingered for a long time just beneath the surface before erupting, allowing time for large crystals to form. Oregon's state gem, the sunstone, is a copper-rich red or yellow feldspar found in one of these lava flows east of Abert Rim.

The Imnaha Basalt

Once the Steens eruptions ceased, new activity began 200 miles to the northeast, near Imnaha, Oregon. The Imnaha basalts (16.6 to 16.1 Ma) erupted from vents now found in the Wallowa Mountains, Hells Canyon, and western Idaho.

The dikes exposed on the steep, uplifted east face of Steens Mountain cut across the horizontal basalt flows. They represent fissures that allowed fluid basalts to reach the surface and erupt. They are now filled with the basalt that did not erupt, and are more resistant than the surrounding lava flows.

The Imnaha basalts are exposed along the canyon of the Imnaha River in eastern Oregon, with the Wallowa Mountains in the distance.

The topography here was relatively rugged, and the resulting basalt flows were confined to stream valleys in easternmost Oregon and Idaho, and small areas of southeastern Washington. They are localized to the canyons of the Imnaha River, and along the Snake River and Grande Ronde River canyons in southeast Washington. A few more ambitious flows ventured as far as Dale, Oregon, 50 miles north of John Day, and may have followed an ancient river channel. Overall, at least twenty-six large flows of Imnaha Basalt erupted, covering 12,000 square miles in basalt lava up to 600 feet thick. This amounted to about 5 percent of all CRB.

The Imnaha basalts present an overall coarse texture, rounded appearance in outcrop, and dark orangey-red-brown color that provides a distinctive appearance for weathered lavas. Like many later groups of the CRB flows, the Imnaha basalts melted and assimilated a trivial amount of the surrounding crust. Chemically, the Imnaha basalts are considered "primitive." Deriving almost directly from the mantle, they are rich in magnesium and contain less silica and potassium than the later Grande Ronde Basalt.

The Grande Ronde Basalt: The Great Floods

The Grande Ronde Basalt is named after the Grande Ronde River in northeastern Oregon, where the formation is superbly exposed. These are the most voluminous and most rapidly erupted of the CRB flows, accounting for 72 percent of the entire CRB. Precision dating by T. L. Barry, along with work by Steve Self and Peter Hooper, suggests that most of the Grande Ronde's 36,000 cubic miles of basaltic lava may have erupted in a period of about 420,000 years—and perhaps in as little as 250,000 years. These are, as Steve Reidel points out, the largest known basalt flows on Earth, with individual volumes of the greatest flows exceeding 2,300 cubic miles.

Like the Imnaha flows, the Grande Ronde lavas erupted from long fissures in the Chief Joseph dike swarm. The massive Grande Ronde flows followed the Columbia River drainage and ponded in the central Columbia Basin. So thick and heavy was the accumulation of Grande Ronde Basalt that the Earth's

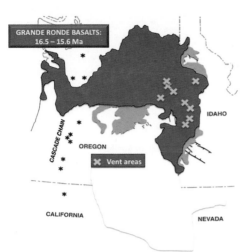

The Grande Ronde basalts were the most extensive of all Columbia River basalt eruptive phases.

crust bent downward beneath what is now the Tri-Cities area, creating a low area where more basalt flows ponded. Today, the total accumulation of CRB flows, including the Grande Ronde and other flows, beneath Hanford, Pasco, Richland, and Kennewick, is more than 15,000 feet thick.

Despite their great volume, the Grande Ronde flows are relatively homogeneous. Compositionally they are not true basalts. Instead, about 57 percent of their weight is silica and they are technically classified as basaltic andesites. However, like the earlier Imnaha and Steens basalts, the Grande Ronde Basalt has a mantle source. As a rule Grande Ronde flows contain few visible minerals. This makes them easy to distinguish from the Imnaha basalts, which commonly contain

The Imnaha basalts were the first of the CRB erupted in northeastern Oregon and western Idaho. They have a more restricted distribution than subsequent phases of the CRB.

Grande Ronde basalts are the most voluminous and widespread of the major flows. Here, they form the walls of the Clearwater River Canyon, Idaho.

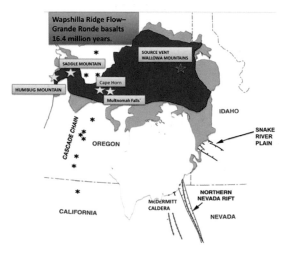
The Wapshilla Ridge flow was the most voluminous and extensive of all CRB flows. It is exposed as the second-from-the-top basalt flow at Multnomah Falls.

Multnomah Falls pours over four Grande Ronde basalt flows, including the Wapshilla flow that erupted in the Wallowa Mountains.

large plagioclase and pyroxene crystals. The relatively high proportion of silica in Grande Ronde lavas indicates that they assimilated substantial amounts of rocks of the crust—quite possibly the roots of the Idaho and Wallowa batholiths, according to John Wolff of Washington State University.

Eruptions of individual large flows persisted for years. Near Maxwell Lake in the Wallowa Mountains, Heather Petrovic and Anita Grunder of Oregon State University have documented eruption rates and volumes of the Wapshilla Ridge flow. This is the largest Grande Ronde flow, as well as the most voluminous basalt flow on Earth. This vent disgorged about 9,000 cubic miles of basalt at an average rate of about 10 cubic miles per day. Eruptive activity may have migrated from one area to another along this 60-mile-long vent during the 3 to 4 years of continuous basalt eruptions, according to Petrovic and Grunder's calculations.

The Wapshilla Ridge flow covered most of the Columbia Basin, extending north to Spokane, south almost to John Day, east to Kooskia, Idaho, and west to the Pacific Coast. It slathered 15,000 square miles in lava. In the Columbia River Gorge, the Wapshilla Ridge flow is the base of Multnomah Falls. On the coast, the Wapshilla Ridge flow forms much of Humbug Mountain, 15 miles east of Tillamook, and is the resistant rock responsible for Fish Hawk Falls near Astoria.

You can find Grande Ronde basalts throughout the Columbia Basin and

In the Columbia River Gorge, Grande Ronde flows form most of the gorge's step walls near and at Multnomah Falls.

Washington's coast, as well as in Oregon and Idaho. They dominate the walls of the Grande Ronde River; are exposed along the Clearwater River in Idaho, as well as the Snake River, especially upstream from Lewiston; and compose the hills south of Pendleton, and most of the Cabbage Hill grade on I-84. Along I-82, north of Yakima, Grande Ronde flows form the base of Umtanum Ridge. They appear in hills north and west of Ephrata. In the Columbia River Gorge, the Grande Ronde basalts form the cliffs at Cape Horn, Multnomah Falls, Horsetail Falls, and the lower cataract of Latourell Falls, where the lava flow filled a small valley, producing the splaying columnar joints. Grande Ronde flows make up the basaltic bluffs west of Rowena, in the area of the Syncline, and much of Dog Mountain as well.

The Prineville and Picture Gorge Basalts

Two important formations of the Columbia River Basalt Group erupted during the Grande Ronde heyday. They are distinctly different from Grande Ronde Basalt, and are formally recognized as members of the CRBG.

The Prineville basalts, or Prineville Basalt Member (15.7 Ma), erupted from vents in a small area south of Prineville, Oregon. They are well exposed at Prineville Reservoir, especially near Bowman Dam, where they look like any other CRB, though with more disheveled columnar jointing, and occasional pillows. The Prineville flows are generally dark brown to black on fresh surfaces, and display few visible crystals. Chemically, they contain higher proportions of phosphorus and barium. Because their chemical compositions are different from other Columbia River basalts, the Prineville basalts have earned their status as one of the seven formations within the CRBG.

The Picture Gorge basalts, like the Prineville basalts, are limited in extent, and erupted at the same time as the Grande Ronde Basalt. The age of these lavas is estimated at about 16.1 to 16 million years. The flows cover 4,100 square miles, mostly in the John Day Basin of east-central Oregon. Altogether, sixty-one flows have been identified, making up about 1.4 percent of all CRB. The Picture Gorge flows have their own compositional peculiarity. They are "high aluminum tholeiites," and according to Peter Hooper, contain the chemical fingerprints of crustal extension and the Basin and Range.

Forests Flourish as Eruptions Take a (Brief) Vacation

The paroxysmal eruptions of the Grande Ronde basalts ended 15.6 million years ago, leaving a somber, level landscape. Cascade volcanoes, humiliated by the enormity of basalt eruptions, had kept their heads down. There is a minimalist volcanic record of their eruptions. However, during the 100,000- to 300,000-year hiatus between the Grande Ronde and Wanapum basalts, the Cascades grew more active, shedding ash and lahars (hot debris flows) east into the lakes and streams that drained toward the ancestral Columbia. The Ellensburg Formation, west of Yakima along Hwy 12, records this formation in its light-colored, layered cliffs that include as many as fifteen lahars from the Cascades.

During this long (to us) interval between major basalt eruptions, the Columbia River and its tributaries formed broad floodplains across the flat landscape. The river's deposits are named the Vantage Horizon—which is considered part of the Ellensburg Formation. Named for marshy sediments, ashy sandstones, lahar and tuff deposits, and mica-rich river sands found at Ginkgo Petrified Forest State Park, a 7,470-acre park just off I-90 at Vantage, Washington, this layer defines the end of Grande Ronde flows across the Northwest. Here, the Vantage Horizon may be hundreds of feet thick. In southcentral Washington it is several feet thick. In the eastern Columbia River Gorge, it is measured in inches.

During this fleeting interlude from volcanic activity, forests prospered and swamps flourished. Some species of the petrified trees preserved in multicolored lithic glory in the Vantage Horizon's layers (maple, oak, walnut, alder, dogwood, spruce, pine) would be familiar—but perhaps seem oddly out of place—in the twenty-first century's xeric and sometimes extreme climate. Others (bald cypress, sequoia, ginkgo, and witch hazel) seem downright anomalous to this place today. How did this diverse assembly of trees—spruce and pine from high-elevation forests, maple and oak from sunny slopes and valleys, and sequoia and bald cypress from torpid wetlands—all come to be jumbled together in one place? A close look at the ash-laden conglomerate they are buried in suggests that the spruce, pines, maples, and oaks might have been entrained in a powerful volcanic mudflow that swept from a Cascade volcano into a metasequoia-lined lake along the ancestral Columbia drainage, mixing alpine, mid-elevation, and riparian species together.

Across the Pacific Northwest, other fossil localities preserve similar Middle Miocene plant communities. At Yakima Canyon, Washington, the seeds, needles, and leaves of a 15.6-million-year-old forest in the Ellensburg Formation included ancestral pine, beech, sweet gum, and Virginia chain fern. Near Grand Coulee, ashy lakebeds between basalts yield leaves of ginkgo, walnut, liquidambar, beech, magnolia, oak, and chestnut. In the remote southeast corner of Oregon, the Succor Creek flora, contemporary with Yakima Canyon, preserves an even more diverse forest of avocado, persimmon, magnolia, tree of heaven, coffee, sycamore, elm, oak, walnut, pear, chestnut, and maple. Conifers included sequoia, pine, spruce, cedar, and white fir. Some of this diversity may be related to differences in elevation, where conifers may have grown on chilly mountaintops and magnolias were relegated to warmer valleys.

Analysis of the mid-Miocene climate in the interior Northwest based on the temperature and precipitation preferences of the ecstatically diverse "mesophytic" forests indicates a warm temperate climate, perhaps similar to southern Arkansas or northern Louisiana. It was, as paleobotanist Kathleen Pigg points out, the Northwest's last stand of a forest type now found only in the American Southeast and in parts of China.

Not surprisingly, in the Middle Miocene, as CRB eruptions peaked, a spike in global temperatures may have contributed to the riotously diverse forests. Known as the Middle Miocene Climatic Optimum (MMCO), the global thermometer rose 5 to 7 degrees F. Based upon soils found between CRB flows, as well as evidence from fossil leaves and ecosystems in general, this gentle warming coincided with the eruptions of Grande Ronde basalts. Higher atmospheric carbon dioxide—the gaseous legacy of Grande Ronde eruption was likely the cause. Grande Ronde eruptions were, like most basalts, propelled by CO_2.

The warmth of the MMCO was followed by cooling, 14.8 to 14.2 million years ago, driven by production of abundant cold, deep Antarctic waters. Like any significant climate shift, the Miocene warming-cooling also generated a global extinction "event," erasing poorly adapted species, especially reptiles, including some species of crocodiles, turtles, and lizards. But it was a rapturous time for plants that could take root in a wetland or ash-laden soil amid an otherwise harsh landscape.

The Wanapum Basalts: Eruptions Resume

Eruptions of the CRB flows resumed with a vengeance about 15.6 million years ago. The new eruptive pulse, known as the Wanapum Basalt Formation (or conversationally, the Wanapum Basalt, would produce much less overall volume, and erupt less frequently, but still deliver many notable basalt flows. This phase tended to grandstand, producing large, long-lasting eruptions, and then quieting for thousands of years. Thin soils and/or sedimentary interbeds separate most flows, indicating that a significant time elapsed between eruptions.

The Wanapum Basalt consists of at least thirty-six flows, but makes up only about 5 percent of the total CRB by volume. Unlike the older Grande Ronde flows, some Wanapum flows (especially the Roza and Ginkgo Members) contain isolated large plagioclase feldspar crystals. Geochemical analyses reveal that the Wanapum lavas assimilated some lower crustal material en route to the surface. However, the Wanapum Basalt is more "primitive" than the Grande Ronde—meaning that they assimilated less crustal material than the more voluminous Grande Ronde flows did.

The three most noteworthy members of the Wanapum basalts are the Ginkgo, Roza, and the Priest Rapids flows. They rank among the ten greatest basalt eruptions in the Northwest.

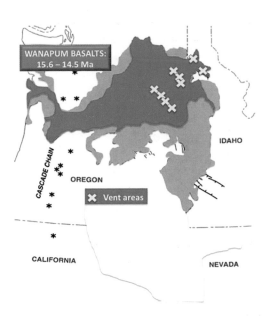

The Wanapum basalts included large and complex flows, including the Priest Rapids and Roza flows.

The Ginkgo Member

The Ginkgo (15.56 Ma) was the first truly large flow to emerge after the Grande Ronde flows ended. The Ginkgo is recognizable by the presence of relatively large and isolated plagioclase crystals, and also its occurrence almost immediately above the Vantage Horizon. It erupted from a 30- to 40-mile-long vent system centered near Kahlotus, Washington, and covered more than 14,000 square miles—an estimated volume of about 380 cubic miles.

The Ginkgo flow extends from southcentral Washington to the Oregon coast, burying 14,00 square miles beneath basalt. At Vantage,

Washington, the base of the Ginkgo flow is a pillow lava complex up to 60 feet thick, indicating that the lava flowed into a sizeable lake here. The Ginkgo's greatest thickness—about 210 feet—is found in the Pasco Basin. It forms cliffs along the Clackamas River near Molalla. It underlies the Willamette Valley. Most famously, the Ginkgo forms much of Yaquina Head and Haystack Rock, 400 miles from its source near Kahlotus, Washington. Anita Ho and Kathy Cashman of the University of Oregon have calculated that this long-traveled basalt flow's average temperature, while flowing lava, was about 2,000 degrees F. It likely moved at velocities as fast as 20 mph under a warm, protective crust of solidified lava, and may have taken as few as 6 days to reach the coast.

The Roza Member

The Roza flows, or Roza Member, about 15.2 million years old, overlie the Ginkgo. The Roza is actually a complex set of four lava flows. Altogether, the Roza covers about 15,500 square miles with 280 cubic miles of basalt. All the flows—or separate pulses of the same flow—erupted from a 100-mile-long vent system in southeastern Washington and northeastern Oregon. Vents and feeder dikes are well exposed near Mayview, along Wawawai Grade Road, and at Potters Hill about 30 miles southwest of Lewiston, Idaho. Stephen Self calculated that the four multiple individual flows that compose the Roza flow field required between 5 and 50 months to get where they were going. With year-long breaks between emergence of each of the major flows, the entirety of the Roza flow system may have required more than a decade to complete its eruptions and solidify.

Like the underlying Ginkgo flow, the Roza flows can be recognized by the isolated large plagioclase crystals. Another endearing characteristic of the Roza is its huge, picture-postcard columnar joints. At Frenchman Coulee, a favorite place for rock climbers near George, Washington, massive columnar joints are up to eight feet wide. The Roza is also exposed near The Dalles in Columbia Hills State Park, Washington, and forms the foundation for much of The Dalles Dam. Unlike the Ginkgo, the Roza flows did not make it to the coast, flowing only to the Bingen-Rowena area.

Many CRB flows erupted vast quantities of carbon dioxide and water vapor, whereas the Roza flows specialized in more corrosive sulfur and hydrochloric acid. Thor Thordarsen of the University of Iceland, and Steve Self at the University of Hawai'i, estimate that the eruptions unleashed about 13,000 metric tons of sulfur dioxide (the rotten-egg smell). During the estimated ten years of its eruption, 700 tons of hydrochloric acid, and 1,800 metric tons of hydrofluoric acid, were also injected into the atmosphere during the estimated ten years of its eruption. This amount of aerosol material, placed as high as 7 miles into the atmosphere, may have initiated a short period of global cooling, and potentially, a decade of "nuclear winter," according to Self. The Roza may not have been unique in its toxic mix. Other basalts, especially larger eruptions,

likely spewed significant amounts of sun-blocking, acid aerosols into the atmosphere. But any chance for substantive, long-term global cooling would have been offset by the high emissions of carbon dioxide and water vapor greenhouse gasses that these basalt eruptions emitted.

The Priest Rapids Member

The Priest Rapids flow erupted about 15 million years ago from vents in the center of Idaho's panhandle near the community of Troy. It is only slightly younger than the Roza flow. Priest Rapids has few apparent phenocrysts—small olivine crystals may be visible in some parts of the flow. By the time the Priest Rapids basalt erupted, the channel of the Columbia River and its tributaries had been severely disrupted. Instead of well-defined channels, small channels crisscrossed the regional drainage system. Lakes developed when lava flows dammed streams. The main Columbia River channel lay far south of today's gorge, cutting through the present location of Bull Run Reservoir and skirting the northern flank of today's Mount Hood. However, 14.5 million years ago, neither Mount Hood nor Bull Run Reservoir were even thought of, let alone present. As the Priest Rapids flow advanced across the flat, sodden landscape, it filled water bodies varying from ponds to large lakes, as well as river channels with hot, steaming lava. Consequently, pillow basalts are common in the Priest Rapids flow. This includes a spectacular outcrop near the junction of I-84

and US 197 at The Dalles, and much of the bottom half of Crown Point where the first of two Priest Rapids lava flows met a lake. The bottom third of Crown Point (about 180 feet) is composed of pillows and breccias formed during that encounter. The upper portion is 300 feet of columnar basalts of ensuing lava that flowed across the now-filled-in lake.

Crown Point marks the western end of the Priest Rapids flow. The huge volume of lava blocked the existing, and probably diffuse, Columbia River channel, forcing the river north. This new channel is known as the

Near The Dalles, Oregon, the Priest Rapids flow entered a shallow lake, producing elongated pillows and lava tubes. Steam produced as the hot lava boiled the lake water altered ash-sized fragments of glassy basalt to clays, producing the yellow matrix of fine-grained palagonite. As the basalt filled the lake, the remainder of the basalt flow found this newly solid ground and produced the more normal columnar jointing found at the top of the exposure.

The uppermost falls at Silver Falls State Park are Wanapum basalts, including the basalt that supports Upper North Falls.

Bridal Veil channel. It would be filled and disrupted by the Pomona Basalt flow, 12 million years ago, which moved the river into roughly its present course.

You can also find Wanapum basalts, caught in the act of crossing the trans-Cascade lowland, in Oregon's Silver Falls State Park. Balanced on the boundary between the Willamette Valley and the Western Cascades, Silver Falls State Park showcases the geology of neither. Instead, it is a place where basalt flows from eastern Oregon followed an ancestral Columbia River channel, 15.3 to 16 million years ago.

Wanapum basalts are the top three basalt flows of the park's North Falls, each fall stair-stepping from one lava flow down to the next. The million-year hiatus between the Grande Ronde and younger Wanapum Basalt is dramatically exposed in the ceiling of the North Falls amphitheater walkway. Here, a thick layer of soils, developed on the older, Grande Ronde flows, once supported a forest. Then, one day 15.5 million years ago, lavas of the Wanapum Basalt flowed through the forest, setting tall maples and oaks and pines on fire—but also preserving their casts and forms. The ceiling of the grotto behind North Falls reveals the charred forest as holes in the ceiling with dark charcoal where each tree burned.

The Saddle Mountain Basalt: The Last Flows

The Saddle Mountain Basalt, named for Saddle Mountain near Yakima, Washington, is the last phase of the CRB. Their volume was small—only about 1.1 percent of the total CRB—and the rate of eruption slowed significantly, with a million years or more between eruptions. The sporadic lava flows includes the Elephant Mountain Member (9.8 Ma), and the Ice Harbor Member (8.8 Ma). Both of these basalt formations are exposed along the Columbia River at Wallula Gap. The last major outpouring of CRB occurred as the Lower Monumental Member (6.2 Ma), which today is exposed at the top of the Snake River Canyon's walls at Lower Monumental Dam.

The Saddle Mountain basalts are the most compositionally and spatially diverse group of CRB flows. Some were relatively rich in silica, some more mafic. Some traveled short distances, others long. The Pomona Member erupted rather leisurely at about 10.95 million years ago in west-central Idaho, and flowed all the way to Radar Ridge and Pack Sack Lookout on the Washington coast, a distance of about 350 miles (!!). Although "tiny" in volume (160 cubic miles, which would cover the city of Seattle beneath a mile of basalt) compared with great flows like the Wapshilla Ridge, Ginkgo, and Roza, the Pomona is the longest documented lava flow on the planet (although one other, 65.9 million years old,

The Pomona Flow is among the last great Columbia River basalt flows, and also the longest. It is best seen in outcrops near Mosier, Oregon, and at Rowena Crest. The Pomona Flow can be distinguished by the presence of small, irregular columnar jointing, called hackly jointing, from top to bottom of the flow.

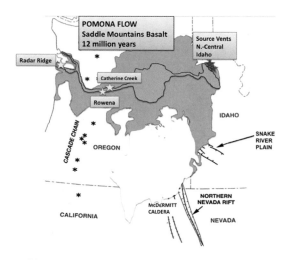

The Pomona Flow, erupted at 12 million years ago, was the longest of the CRB flows, and the longest lava flow known on Earth.

in India, rivals and may exceed it in length). The Pomona flow is distinguished by "hackly" jointing—small irregular columns. It followed the course of the Columbia River, and today is well exposed in cliffs and roadcuts just east of Mosier, Oregon, and at Rowena Crest.

The Shifting Rivers

The Columbia River basalt lavas reached the Pacific coast frequently because they usually followed the Columbia River's broad valley across the Columbia Trans-Arc Lowland to the sea. The Trans-Arc lowland extended from near Clackamas, Oregon, north to the Washington border. It provided a 30-mile-wide, easy pathway across the then-low Cascade Range into western Oregon and the Willamette Valley.

The Columbia has changed its course and configuration significantly during the past 17 million years. When the CRB first erupted, the river likely emptied into the sea near the present site of Newport, Oregon. The course changed as basalt choked the river's channel, forcing the stream to move and carve a new channel, or detour around the lava flows. Steve Reidel and Terry Tolan documented the ensuing dance between river, lava flows, and tectonics, noting: "The peak of flood basalt volcanism obliterated river paths, but as flood basalt volcanism waned, the rivers were able to establish courses within the growing (Yakima) fold belt. As folds grew larger (about 14–10 million years ago), the major pathways of rivers moved toward the center of the Columbia Basin where subsidence was greatest. The finishing touches of the river system,

however, were added during the Pleistocene by the Missoula floods, which caused local repositioning of the river channels."

The larger Columbia River system pre-dates the CRB flows. But the rivers' exact configuration is unknown. We do know that the Salmon and Clearwater Rivers provided the principal drainage for the Northwest interior, and likely joined an ancestral Columbia near the center of the Columbia Basin, possibly near today's Moses Lake. Until about 4 million years ago (the Pliocene), the Snake River followed a different path, probably flowing through the Grande Ronde Valley and then northwest toward Umatilla, with a confluence with the Columbia/Clearwater system near today's The Dalles, Oregon, about 16.5 million years ago.

By 16 million years ago, the vast Grande Ronde flows flooded the Columbia Basin. These huge lava flows blocked drainages through the basin and displaced the Columbia River's course north and westward, so that the river flowed along the eastern flanks of the Cascades. The Columbia's course across central Washington was reduced to a series of shallow lakes, as well as streams that wandered across a crenulated basaltic landscape. Once it reached the present-day site of the town of Hood River, the Columbia River's main channel sagged south almost to the present site of Eugene. The Snake and Clearwater River systems joined the Columbia about where Mount Hood rises today. Reidel and Tolan suggest that by 15.6 million years ago, the Columbia's main channel was forced back into the central Columbia Basin by the uplift of Naneum Ridge, near the present site of Priest Rapids Dam. The Columbia River finally occupied a channel that approximates today's course by about 12 million years ago.

Invasive Flows: The Columbia River Basalts on the Coast

When the Columbia River basalt flows reached the coast, they behaved very differently than they had throughout the rest of their journey. The Oregon and southwestern Washington coast lay 15 to 20 miles east of its present location. The Columbia flowed into a wide embayment. Sands that today compose the Astoria Formation were an offshore fan.

Saddle Mountain, one of the highest and most distinctive peaks in Oregon's Coast Range, is composed of Grande Ronde basalts that built a lava delta as they burrowed seaward beneath sediments of the Astoria Formation.

They were soft and relatively unconsolidated. As lava flows encountered the Pacific, they built a "basaltic lava delta" atop the Columbia River's sedimentary delta. Many flows sank into the soft, unconsolidated sediment as they entered the water.

Hot lava and cold seawater are an explosive mixture. Explosions, boiling steam, and swirling currents often fragmented the lava. The result is a broken rock, re-cemented and stuck together while the fragmented lava was still hot. If molten basalt is able to flow under water, it may also form pillows. Pockets of still-molten basalt squeezed into cracks and continued upward, forming dikes within the broken-up rock. Saddle Mountain provides excellent examples of this process. It is an erosional remnant of the delta, its rocks formed largely under water. Saddle Mountain is composed of fragmented basalt called breccia or, if the fragments are small, peperite. Pillow breccia—broken basalt and interbedded pillow lavas—are also present. Narrow basalt dikes—lava squeezed downward through the broken and loosely cemented basalt—appear as streaks on Saddle Mountain's south side.

Many of the flows that reached the coast were among the most voluminous and extensive basalt flows ever unleashed on the planet. Lava continued to flow over and through the porous lava deltas for months, filtering through them in conduits or dikes that led into the subsurface of the loose Astoria offshore sands. This style of emplacement is known as an invasive flow. These are exposed in sandstone sea-cliffs at Arcadia State Wayside, Humbug Point, and Hug Point State Park.

Sometimes larger pockets of the stealthy subsurface lava approached the size of a small intrusion. Pressurized by incoming lava and the overburden of sand and seawater, these subterranean pools of lava vented or erupted onto the surface. The result is additional pillow lavas, breccias, and peperites. Haystack Rock near Cannon Beach, Oregon, is an example of these effusions of invasive lavas.

Haystack Rock is a remnant of the Ginkgo flow that burrowed into the subsurface and then worked its way back to the surface to emerge as an offshore eruption. The thinly bedded mudstones that the basalt erupted through suggest that the water was quite deep at the time of Haystack Rock's emergence. On close examination, Haystack Rock is composed of pillow basalts, formed as the lava erupted from the sands; brecciated, broken, and fragmented basalt that is fused together; and several solid invasive sills that formed as hot lava intruded the now-solid but fragmented basalt.

In Oregon, invasive sills underlie many coastal waterfalls, including Fish Hawk Falls and the 319-foot-high Munson Falls. Remnants of large subsurface lava reservoirs can be identified by the presence of well-developed columnar jointing, and include much of Cape Lookout and the sea-cliffs at the Cape Mears and Tillamook Rock lighthouses. Gravity and magnetic surveys indicate that today, these subsurface basalts continue another 34 miles west of the Tillamook Rock Lighthouse at depths as great as 6,000 feet beneath the sea bottom.

Columbia River basalt lava flows intruded into the soft sands of the Astoria Formation, producing sills and dikes, as well as odd patterns of intrusion. These invasive basalts support most coastal headlands from Yaquina Head northward.

The iconic Haystack Rock in Cannon Beach, Oregon, is composed of the Ginkgo Flow of Wanapum Basalt that worked its way up through overlying sediments to form a small eruptive center on the ocean floor.

The Miocene Coast and the Astoria Formation

If you were to visit the Pacific Northwest's beaches 20 to about 14 million years ago, you would pitch your umbrella about 10 to 15 miles east of where Pacific waves break today.

The dominant Miocene sediments of this time are sandstones deposited as deltas and offshore fans, collectively known as the Astoria Formation. Altogether, the Astoria Formation is about 2,000 feet thick, according to George Moore and others at Oregon State University. Its sands include rounded fragments from eastern Oregon's Blue Mountains as well as the Oregon Coast Range. The Astoria Formation also includes thin coal seams that provide evidence of nearby forests, and fossils of alder and maple leaves.

But marine fossils are the real treasure of the Astoria Formation. They include marine mammals such as whales (dominantly the Mysticete/baleen whale, *Cophocetus oregonenis*), seals, seacows (*Desmostylus*), and occasional dolphins. By far the most abundant and collectable fossils are marine mollusks, including clams, scallops, marine snails (gastropods) with lovely whorled shells, and an occasional nautilus. Most mollusks are species adapted to shallow depths and sandy bottoms. Throughout the Astoria Formation, rock-dwelling organisms such as anemones are absent. Paleontologist Ellen Moore has identified 97 species representing 73 genera and 45 families in the Astoria Formation along the Oregon and Washington coasts. Oregon's Beverly Beach State Park offers easy access to collectible fossiliferous strata, newly exposed each year by the power of the Pacific.

Marine mollusks like this scallop are common fossils in the Astoria Formation sandstones.

The Owyhee Calderas

The Owyhee region of southeastern Oregon and southwestern Idaho is a buckskin-tinted landscape of tawny tuffaceous cliffs and black basaltic rimrocks. It is a sort of geologic nudist colony, where the anatomy of ancient volcanoes, ancient lakes, and ancient life are seldom garbed by vegetation and are visible in their entirety for all to admire.

The volcanoes are the younger and more explosively irresponsible siblings of Steens Mountain. Steens and related basaltic vents in southeast Oregon ceased their calm, basaltic eruptions about 16.2 million years ago, but the heat and magma that remained beneath the surface festered. The source of heat may have been partly due to the Yellowstone hotspot, rather than the ruptured plate. We are not sure. The result was the eruptions of large calderas, along with more demure—but still significant—outpourings of dark, iron-rich basalt or light-colored, silica-rich rhyolite. Chocolate or vanilla. Nothing in-between. Why only two kinds of volcanic rock—or what geologists term "bimodal"? Because sometimes, the hot basalt pushes its way to the surface and erupts into cinder cones and basalt flows. At other times, rising basaltic magma stalls on its way to the surface to erupt, and the heat of the intrusion melts the overlying crust, producing an explosive rhyolite.

In the Owyhees of Oregon and Idaho, multiple calderas erupted between 15.5 and 15.1 million years ago. They include Mahogany Mountain, Three Fingers, Castle Peak, Grassy Mountain (a site once proposed for a gold mine), and Star Mountain in Oregon. Mahogany Mountain is the remnant of the largest and oldest-known Owyhee caldera. About 8 miles wide, it erupted 15.5 million years ago. In Leslie Gulch, a deep canyon incised into tuff on the north flank of Mahogany Mountain, ash-flow tuffs (Leslie Gulch Tuff), explosion breccias, and thick air-fall tuffs compose most outcrops. The massive outcrops that make up the walls of Leslie Gulch are part of the Mahogany Mountain caldera's central vent complex. The huge holes in many Leslie Gulch Tuff exposures are the legacy of abundant bubbles of steam that were present in the hot but very gaseous and fluffy cotton candy–like tuff. As the molten tuff solidified, the "bubbles" remain. Smaller, related eruptions built small pyroclastic cones, including the Honeycombs, a vent on the north flank of Mahogany Mountain that spewed frothy tuff and juicy, water-laden rhyolite.

As caldera eruptions waned 15 million years ago, a 30-mile-wide down-faulted basin—the Oregon-Idaho Graben—took shape. Work by Michael Cummings and others has shown that this rift system extends from northern Nevada north almost to Baker City, Oregon. The Oregon-Idaho Graben is part of a larger system of rifted basins that extend along the western edge of the ancient craton and include the La Grande and Baker Grabens to the north, and the Northern Nevada Rift to the south.

Rhyolite lavas, ash-flow tuffs, and basalts erupted within and adjacent to the graben, filling it as it subsided. Among the earliest were the Dinner Creek Tuff (15.3 Ma) and the older Hunter Creek Basalt (15.9 Ma), which form extensive rimrocks west and southwest

The Owyhees of southeast Oregon and southwest Idaho expose the remains of explosive Miocene calderas that erupted during and somewhat after the Columbia River basalts. They consist of light-colored, silica-rich rhyolites and tuffs, as well as dikes that fed the eruptions, including these exposures in Leslie Gulch.

of Vale, Oregon. Later eruptions include the Jump Creek rhyolite (11.5 Ma)—a stunning, rose-tinted rhyolite flow up to 800 feet thick exposed along Succor Creek at Succor Creek State Park. East of Succor Creek, the Pole Creek rhyolite (10.6 Ma) forms a high-rising plateau up to 900 feet thick. Far to the north, near Baker City, rhyolites also formed Dooley Mountain (11 Ma), and perhaps the Strawberry Volcanic Field (14 Ma) close to John Day. Basalts also leaked to the surface along the graben's faulted edges.

The last gasps of Owyhee activity about 12 million years ago include basaltic shield volcanoes of Freezout Mountain and Cedar Mountain, rhyolites at Double Mountain, and basalt flows on the flanks of Grassy Mountain. Small amounts of younger volcanic rocks, including the basalts of Jordan Craters, less than 10,000 years old, dot the Owyhee region's landscape. These young rocks are related to the High Lava Plains volcanics, rather than the older Owyhees.

Life of the Middle Miocene

The Miocene landscape of the Northwest was a basalt-riven grassy plain, dotted with woodlands. At Succor Creek in the Owyhees, volcanic ash has preserved the fossil leaves of a diverse plant community. At least 150 different species represent a variety of environments, including wetlands, dry grassland slopes, and uplands. All plant species identified required strong summer rains. Throughout the region, paleofloras indicate that the topography was lower and wetter in the north, and higher and drier in the south. Plants include ancestral avocado, persimmon, dogwood, coffee tree, magnolia, tree of heaven, sassafras, and walnut, as well as spruce, pine, and white fir. Birch, alder, oak, metasequoia, and willow composed the mid-elevation riparian deciduous forest. Some fossil leaves found in the Succor Creek Basin were so well preserved by rapid dehydration and fixation by organic acids in warm ash that cell walls, chloroplasts, and native, original cellulose are retained.

Extraordinarily well-preserved Miocene plant leaves are also found in 15-million-year-old lakebeds near Clarkia, Idaho. Here, the usual scenario of a stream dammed by a Columbia River basalt flow produced a 400-foot cold, deep, narrow, anoxic lake. Leaves that sank to the bottom were quickly entombed in fine silts and clays, chilled, and deprived of oxygen required for decay. The result was a classic "Lagerstätte"—a deposit where exceptional preservation in the absence of oxygen includes intact soft tissue. The leaves of magnolia (*Magnolia latahensis)* and laurel (*Persea pseudocarolinensis)* yielded sufficient DNA for partial sequencing by Sangtae Kim and colleagues at the University of Florida. The results demonstrated that the magnolia was more closely related to *Liriodendron* (tulip magnolia) than to *M. grandiflora* as presumed, and that Miocene laurels were closely related to sassafras. This forest (and its climate) may have been similar to forests growing today on the eastern side of the Appalachian Mountains in North Carolina, South Carolina, and northern Georgia.

Sod-forming grasses dominated, rather than the bunchgrasses found in the region today, according to Greg Retallack's analysis of Miocene soils preserved between basalt flows in the John Day Fossil Beds of eastern Oregon. The Miocene grasslands were likely composed of *Poa* species, similar to the eighteenth-century Great Plains. Grasses were more adapted to grazers in the Miocene. And grazers grew more adapted to grasslands.

The animals that lived here while the Columbia River basalts erupted must have been a fleet and cautious lot. One day you had grass to eat. The next day, everything was toast.

Horses changed. *Parahippus*, the most forward thinking of the Miocene horse clan, replaced *Miohippus* as the dominant grazer. *Parahippus* stood about 10 hands and had legs with less sideways rotation capacity that were better adapted to running across open ground rather than navigating dense forests. It had one large central hooved toe, with tiny side toes that had almost disappeared. Its cousin, *Merychippus*, was smaller (7 hands) but had teeth exquisitely adapted to grazing on grass, with much wider molars and sharp incisors. It became arguably the most abundant and prolific of the Miocene horse clan. By the end of the Miocene, a 10-hand-high *Neohippus* cantered across the Northwest's dry shrub-steppe grassland and basaltic plateaus, and *Pliohippus*, with long legs, springy pasterns, and hard single hooves, attained true equine dimensions. The Neanderthals of the horse clan seem to have been *Kalobatippus* and the smaller *Archaeohippus*. Both retained their forest-friendly format into the Middle Miocene, perhaps 19 million years ago, including springy multi-toed feet and teeth designed for munching soft forest leaves. As grasses seized the landscape and deciduous forests shrank in the Middle and Late Miocene, *Kalobatippus* and *Archaeohippus* disappeared.

The chalicothere *Moropus* (or, more properly, *M. oregonensis*) was perhaps the oddest animal of the Miocene's menagerie. If a committee designed an "average large mammal" this is what they might have come up with. Standing 7 to 8 feet tall and weighing an estimated 250 to 300 pounds, it had the massive looking body of a bear with three fearsome long claws on its front feet similar to a giant sloth. Despite this ferocious appearance, it was a peaceful, plant-eating perissodactyl, with the head of a horse, including teeth designed for both cutting and grinding leaves. The claws, evidently, were used for stripping leaves when it stood in an upright posture, and perhaps were handy for digging up succulent roots as well. It had no tail. In the Northwest, chalicotheres are known mostly from scattered remains—a tooth here, a toe there—found in and around the John Day Fossil Beds. These remnants are bracketed by ashy rocks dated at about 21 million years in age—pre-dating the Columbia River basalt eruptions. Overall, chalicotheres roamed North America for all of the Miocene, appearing globally about 25 million years ago and disappearing only 5 million years ago.

Several species of rhinos roamed the Miocene's partly open, partly forested, and basalt-laden landscape. The most intriguing of their remnants are known as the Blue Lake Rhino—the mold of a rhino's body that was caught and entombed in CRB lavas near Blue Lake, Washington, south of Grande Coulee. A "tramping party" of hikers discovered the mold in a small lava cave on the west side of Blue Lake in 1935. Investigated in detail by Warren Chappell and others in 1936, this animal measured about 94 inches in overall length, and, based on its form, with stubby legs outstretched, stood about 4 feet high. Bones, including an intact jawbone replete with teeth, suggest the animal was a *Diceratherium*, a relatively small Miocene rhino. Males had two horns side-by-side on their nose rather than one horn as modern rhinos do. Females were hornless. The Blue Lake rhino, Chappell concluded, was mercifully already dead when lavas overtook it. "The animal was lying on its left side and its short legs are in a position characteristic of an animal in rigor mortis," he observed. "The body mold resembles that of an animal of heavy build. The head is long and somewhat pointed." Chappell, ever the storyteller (as are most geologists) sets the scene of the rhino's preservation in his 1951 paper. "Apparently the advancing lava flow encountered a shallow body of water," he writes, "with a dead and bloated rhinoceros either upon shore or floating on the water. The water caused the formation of pillows at the base of the flow. The pillows retained sufficient plasticity to pack around the body of the animal

but were rapidly cooled by water so that they became rigid enough to preserve the mold thus formed." What led to the Blue Lake rhino's death in the first place? Perhaps fright, or running in terror from ceaseless, remorseless, torrid, fast-moving, and overwhelming lavas.

Overall, the Middle Miocene, with its higher temperatures and islands of grasslands and lush forests, provided exquisite habitat for a great diversity of animals besides horses, chalicotheres, and rhinos. Camels—small, large, and giraffe-necked—trekked across open spaces. Small shovel-tusked elephants called gomphotheres rummaged through forests. Multiple species of deer and antelope roamed. True dogs, including the bone-crushing *Tomarctus*, began to replace bear dogs. True cats appeared, complete with saber teeth.

Mid-Miocene warmth, partly supported by Columbia River basalt flows and enhanced atmospheric carbon dioxide, came to an end about 14 million years ago, as basalt eruptions waned. Plate tectonic motions forced changes in ocean circulation, and orbital factors (Milankovitch cycles) reduced the solar influx. Known as the Miocene Disruption, this event dropped global temperatures rapidly. In the Northwest, climates became cooler and drier. Shrub-steppe replaced forests and lush grasslands. Some species of horses, bear-dogs, and antelope-like Dromomerycidae faded away, producing a small extinction event 14.5 to 13.5 million years ago. The next chapter in the Northwest's history would see the rise of mountains and the development of true deserts, before the Ice Age began.

Winter Rim and Summer Lake, Northern Basin and Range, Oregon. Faulting continues to uplift the rim and tilt Summer Lake to the north by about 2 cm since the late Pleistocene. Features named by John C. Freemont, whose expedition slogged through hip-deep snow at the rim—and found an inviting green, water-rich basin, complete with hot springs, spread out in the Summer Lake basin below.

Global Temperatures

Millions of years | 2500 | PROTEROZOIC | 541 | PALEOZOIC | 252 | MESOZOIC | 66.0 | CENOZOIC

Camb | Ordo | Silur | Dev | Miss | Penn | Perm | Trias | Juras | Creta | Paleo | Eoce | Olig | Mioc | Plioc | Pleist | Holo | Anth

CHAPTER 9   The Great Expansion   *The Basin and Range Opens*

The Pacific Northwest hosts the northern portion of the Basin and Range province—a region of unabashed desert and alpine forests that include most of Nevada and snippets of Utah and California as well as Oregon. In Oregon, abrupt mountains and flat, barren playas occupy Oregon's southeastern quadrant, extending from Klamath Falls on the selvage of the Cascades, to Rome, Oregon, where the Owyhees begin. The geography comes in pairs. Steens Mountain and the Alvord Desert. Hart Mountain and Warner Lakes. Abert Rim and Abert Lake. Winter Rim and Summer Lake. As John McPhee noted: Basin. Range. Basin. Range. Basin. Range.

This is a landscape crafted by faulting during the past 12 million years. The faults that uplift ranges and down-drop basins are "normal faults"—displacements that allow one side of the fault to slip down and away from the other. More specifically, they are "listric normal faults." The term "listric" comes from the Greek word *listron*, meaning shovel or spoon. A listric fault is curved—steep near the surface, but significantly flatter at depth. This process allows blocks to rotate, uplifting one part of the block, while down-dropping and extending the rest. The ranges of the Basin and Range are rotated and tilted as well as faulted. Most major Basin and Range faults have impressive offsets. The rim of Steens Mountain towers more than a mile above its original position, which was somewhat below the present Alvord Desert floor. This displacement along the 40-mile-long Alvord Fault system has taken more than 12 million years—a vast time of moving 1/2 inch every millennia or so.

The Basin and Range is expanding much faster horizontally than vertically. Nevada is getting wider at the rate of about 1 inch per year, a similar rate to oceanic crust. Over the past 12 million years, the crust of Nevada and southern Oregon has managed approximately 100 miles of expansion westward.

Hot Springs, Energy, and Faulting

This extension thins the crust as it extends the landscape. Today, the crust in most of the Northwest provides a blanket of insulation between the mantle and us that is 19 to 60 miles thick. In the Basin and Range, much of the crust is a fragile, faulted, and fractured 15 to 18 miles thin. Its heat lies just beneath us. It is no wonder that hot springs accompany most Basin and Range fault systems. In the Alvord Desert, hot springs that include Borax Lake and Alvord Hot Springs surface along the Alvord Fault at the base of Steens Mountain and the Pueblo Mountains. Alvord Hot Springs are the best known and most accessible. The natural hot springs emerge at a temperature of 180 to 170 degrees F. Hot waters are piped to a soaking pool that maintains temperatures of about 110 to 115 degrees F. Borax Lake, about 4 miles east of Fields, Oregon, is also a hot spring site that includes a 10-acre lake. The water level in Borax Lake is higher than the surrounding desert floor. The higher elevation is attributed to deflation (wind-erosion) of the desert floor adjacent to the lake, while lake waters cemented the lake banks into a more resistant material than the surrounding sands.

At the base of Hart Mountain, near Plush, Oregon, Antelope Hot Springs and Fisher Hot Springs rise along a system of active faults. In Nevada, geothermal energy powered by mantle-driven heat is a promising source of electricity, with projects that include the 47 megawatt Stillwater 2 Plant, and the 51 megawatt Dixie Meadows Plant that will come online in 2015.

Oregon's Basin and Range is seismically active. In 1996, two magnitude 6.0 earthquakes shook Klamath Falls, on the western edge of the Basin and Range, killing two and causing $12 million in structural damages. The new Klamath County Courthouse and the new regional library both replace buildings mortally wounded by these temblors and the thousands of aftershocks. Other swarms of quakes, and individual events, have rattled the region, including the 1976 swarm of thousands of small earthquakes from magnitude 4.5 to 1.0 in Warner Valley north of Adel, Oregon.

Rearranging Northwest Landscapes

Basin and Range extension has radically rearranged West Coast geography, 1 or 2 inches at a time. The relentless pressure from the expanding crust has produced the folds of the Yakima Fold belt, as well as uplifted the Blue Mountains. Faulting to accommodate this movement occurs on the Brothers Fault Zone and Olympic-Wallowa Lineament and allows the Sierra Nevada, Klamaths, and Coast Range to rotate westward like the hour hand of a giant clock moving from the five o'clock position to the eight o'clock position.

The Basin and Range's expansion has displaced much of the geography of Oregon and Washington (and California) westward. It is blamed for the separation of the old rocks of the Klamaths from potentially related terranes in the Blue Mountains. It also makes the neat north-south alignment of the Cascades a modern, fortuitous accident.

This rotation is very active today. The Cascades and Coast Range, and even the Sierra Nevada, are swinging west around a pivot point. The hinge-pin for expansion is varyingly located near Seattle, or near Maupin,

Hot springs bring scalding, acid waters to the surface of the Alvord Desert in the Nature Conservancy's Borax Lake Reserve.

The Alvord Desert is typical of arid basins, with hot springs and an annual rainfall of about 7 inches.

Looking up the Columbia River toward Mosier, the down-fold of the Mosier syncline is evident. The Syncline was formed as the Basin and Range has expanded, and tectonic rotation exerts pressure on eastern Oregon and Washington.

Oregon, where small, barely noticed quakes centered in the same mile-square area rattle the crust weekly. The Basin and Range pushes at the underbelly of the Northwest, shoving up the folds of Central Washington, shaking the faults of southcentral Oregon, and driving stresses on faults in Portland and Seattle.

The Mechanism of Expansion

The source of the Basin and Range's expansive unrest is a puzzle. One line of evidence—the frequent eruption of basalts through the thinning crust—suggests a mantle plume. But if it is a plume, it seems to move west at the same rate as the continent, unlike other plumes that are fixed as the continents move above them. Another line of reasoning suggests that the subducted Farallon mid-ocean ridge—overrun by North America about 30 million years ago—simply could not turn itself off, and kept spreading as it went down. This idea, too, has

many flaws. Including how to explain why a subducted, rootless mid-ocean ridge would continue operation.

The newest and most plausible idea is a process called delamination. The idea is simpler than it sounds. First, a thick accumulation of rocks piles up on a continental edge. This occurred during the Jurassic through Cretaceous, 140 to about 90 million years ago, as exotic terranes were shoved onto the west coast. Then, this newly acquired pile of terranes settles deeper into the mantle. And as it does, the bottom of the pile experiences great pressure. Obeying basic laws of chemistry and physics, the rocks under high pressure become denser. This means they are not well supported by the surrounding, more fluid, upper mantle and asthenosphere. These denser rocks break off and sink farther into the mantle. That removes the weighty "anchor" from the lighter rocks that were piled up on

A fault near Weston, Oregon, illustrates the typical direction of off-set on the Basin and Range normal faults, with one side down-dropped relative to the other.

Abert Rim epitomizes Oregon's Basin and Range. It rises abruptly from the desert floor like a giant, looming wall. It is the most formidable and longest fault scarp in Oregon, extending more than 30 miles from its inception near Valley Falls north to Alkali Lake. At its greatest height, Abert Rim towers more than 2,500 feet above salt-ridden Abert Lake, Oregon's most saline lake and a ghost of vast Great Basin and High Desert lakes of the past.

the continental edge, allowing them to rise. Without the dense root to support them and hold them together, they ultimately begin to collapse under their own weight and spread out. The crust begins to extend. This thins the crust, allowing heat and magmas from the mantle much easier access to the surface. Voila—an extending region of crust that slowly, inexorably shoves the adjacent crustal blocks north and west as it flattens out like a mound of cookie dough.

The High Lava Plains and Brothers Fault Zone: A Trolley for the Expanding Crust

The barren landscape of Oregon's High Lava Plains is home to an important but unobtrusive set of faults that allow the Basin and Range to expand: the Brothers Fault Zone (BFZ). This region of short strike-slip faults marks the northern boundary of the Basin and Range. The BFZ extends from the northern end of Steens Mountain to just north of Newberry Volcano. These faults are a sort of "zipper" or "trolley" that accommodate the Basin and Range's westward motion, buffering the region to the north from the stresses of lateral motion. If you drive Hwy 20 from Bend to Brothers to Burns, you will spend 2 hours or more in this faulted landscape. But you will see scant evidence of crustal angst. The BFZ does its work efficiently and well, unleashing multiple temblors of magnitude 1.0 or less every year on several small and parallel faults, allowing the crust of the Klamaths and Sierras and Cascades to move west. Other

parallel fracture zones, including the Eugene-Denio and fault systems to the south, and the Olympic-Wallowa Lineament to the north, may play similar, though less significant roles.

The arid volcanic landscape that accompanies the Brothers Fault Zone is known as the High Lava Plains. This region of faults, vents, and volcanic rocks is bounded by the Blue Mountains on the north, Cascades on the west, Basin and Range to the south, and ancient North American craton (and Idaho) to the east. The crust is thin and fractured here. Volcanic rocks erupted during the past 10 million years abound. Volcanic activity mostly comes in two flavors: basalt or rhyolite "bimodal" volcanism.

The oldest volcanic vents of the High Lava Plains include 10.4-million-year-old Duck Butte at the northern end of the Steens Fault. The youngest is Newberry Volcano south of Bend, which produced one of Oregon's youngest lavas, the Big Obsidian Flow, which erupted about the year AD 800. Intermediate ages belong to Glass Buttes (5.5 Ma) and the flat-lying caldera at Capehart Lake, 30 miles west of Burns, which erupted the Rattlesnake Tuff 7.1 million years ago. They are opposite what one might expect. If these lavas were generated by a Yellowstone-like hotspot, the oldest lavas would be in the west, the youngest in the east. But volcanic activity in the High Lava Plains has moved in the opposite direction. It is youngest in the west (Newberry and South Sister) and oldest in the east (Duck Butte, just north of Steens Mountain). This is the mirror image of the ages along the track of the Yellowstone hotspot along the Snake River Plain.

The causes and sources for this activity are enigmatic, notes volcanologist Mark Ford. It is difficult to explain why volcanoes sprang to life far from the active edges of a tectonic plate. Ford's recent research, along with Anita Grunder and others at Oregon State University, suggests a complex mechanism for the High Lava Plains. First, as the modern Juan De Fuca plate shoves its way slowly into the mantle, it generates ponderous back-flow in the mantle above it. The rising mantle rocks melt a bit as pressure lessens. This produces basalt. Some basalts filter through the fractured and faulted crust and emerge on the surface as small-scale lava flows. But others are trapped beneath the crust of accreted terranes. Basalts are hot. Pools of basalt will gradually melt the overlying crust. This produces a rhyolite lava flow—and sometimes explosive activity. Voilà! Bimodal volcanism. Rube Goldberg would be proud.

But how do you get the older eruptions in the east and younger in the west? Ford's High Lava Plains model—which includes very sophisticated geophysics and geochemistry—lays the blame squarely on the Brother's Fault Zone and the rotation of the Basin and Range.

If you hold a sheet of paper in your hands (You *can* try this at home . . .) and then try to "rotate" the bottom half in a clockwise

The High Desert volcanics include tuffs and ignimbrites such as the Westfall Tuff and several major ash-flow tuffs.

Newberry Volcano is one of the youngest eruptive centers associated with the High Desert Volcanics. The Big Obsidian Flow is about 1,200 years old.

direction (like the Basin and Range extension is producing) you will notice that to accomplish this, the paper must tear apart. The right side (east) rips first. The tear propagates toward the left (west.) This simple experiment demonstrates that the Brother's Fault Zone propagates westward, and that its faults became more numerous and open on the east end before they loosened up in the west. So it makes sense that the first place that thick, sticky rhyolites could make it to the surface and erupt was on the east end of the fault zone (think Duck Butte). On the west, it took much longer for large fractures to develop so that viscous rhyolite lavas could erupt. Think Big Obsidian Flow at Newberry Volcano, or Rock Mesa, near South Sister.

The High Cascades Begin

Faster rates of subduction generated greater activity in the Cascades during the Late Miocene. In Oregon and northern California, volcanoes blossomed. In Washington, upfolding and intrusion was more popular. This new bloom of Cascade volcanoes marks the beginning of the High Cascades, a range that first erupted about 10 million years ago, although the majestic peaks we recognize today are generally less than a million years old.

In northern California and southern Oregon, more rapid subduction increased eruptions, and also pulled the crust apart, producing easy avenues for lavas to reach the surface. Near Alturas, California, east of today's Mounts Shasta and Lassen, basalts of

Hoodoos—statuesque formations produced when a column of soft tuff is sheltered from erosion by a resistant caprock—expose the Deschutes Formation above the Metolius River, Oregon.

the Devils Garden are about 5 million years old. They seeped from the mantle to the surface through thinned and faulted crust, like blood oozing through an abraded wound. Similar lavas formed the base of the Modoc Plateau, the east flank of the Cascades near Klamath Falls. Mount Lassen is surrounded by (and built atop) more than 500 vents that are between 10 and 3 million years old.

Central Oregon's Deschutes Formation reveals sporadic eruptions of ash and ignimbrites, tempered by calmer episodes of fluid

basalts, 7.5 to about 5 million years old. Today, the volcanoes that produced these eruptions are nowhere to be found. This is because crustal extension pulled apart the crest of the Cascades between Three Sisters and Mount Jefferson, down-dropping the Late Miocene vent areas into a broad, faulted valley known as the High Cascades Graben. Subsequent eruptions filled the graben, covering the older volcanoes. However, these older vents are well exposed on Green Ridge, northeast of Sisters, where a layer-cake of lava flows and dikes appears in the steep western face.

Evidence of Late Miocene High Cascade eruptions is easy to find at the base of Mount Hood. It includes the Rhododendron Formation—andesites and tuffs found west and north of Mount Hood. The Laurel Hill pluton, light-colored diorites exposed in rocky roadcuts along Hwy 26 between Rhododendron and Government camp, marks an 11.6-million-year-old magma chamber that fed early Rhododendron eruptions. The location of the actual vent remains unknown, and is likely buried beneath Mount Hood. Ignimbrites and mudflows from this ancestor of Mount Hood traveled north, and today form the white cliffs of The Dalles Formation near The Dalles, Oregon. Andesites of similar age occur along Polallie Creek on the northeast side of Mount Hood.

While much of Oregon and California's ancestral High Cascades were pulled apart, in Washington, they were pushed together—or more correctly, pushed up. In Washington, most Late Miocene volcanics have been uplifted, eroded, and erased. In places, rugged topography of the Cascades is the legacy of erosion, rather than eruption. At the northern base of Mount Rainier, the Tatoosh pluton peeks out from the volcano's flanks. But unlike Mount Hood, which also bares a plethora of Late Miocene volcanic rocks that stretch north to The Dalles, there are few remnants of any volcanic eruptions. Some volcanic roots are there. But the forest of towering

Late Miocene volcanoes have been, so to speak, geologically clear-cut and hauled to the sawmill. Peter Reiners and colleagues of Yale University calculated (based upon helium isotopes in exhumed rocks) that prior to 12 million years ago, uplift in the Washington Cascade crest amounted to a paltry 600 feet per million years (or 0.7 inches per century). However, between 6 and 12 million years ago, the uplift rate accelerated to a nose-bleeding 4 inches per century. Rapid uplift is often associated with greater erosion. So whether the absence of volcanic rocks reflects a dearth of volcanoes or highly efficient erosion is not certain. Generally, the bets are on fewer volcanoes.

The reason for this disparity—uplift and fewer volcanoes in Washington, extension and greater activity in Oregon and California—is unclear. Peter Reiners and colleagues point to a shift in both the direction and speed of plate convergence about 8 million years ago. They also reason that delamination, or peeling away the lower crust from the northwestern Cordillera, while leaving it intact beneath Oregon and Washington, may have helped accelerate uplift. This would increase erosion rates, erasing much of the Washington Cascades. Evidence suggests that at least part of Washington's lower crust is missing. Beneath the Washington Cascade Range and much of British Columbia, the crust is only 25 to 30 miles thick, with no sign of a mafic root. Beneath much of Oregon, the crust is a more comfortable and conventional 40 miles thick.

The Late Miocene was another time of climatic adjustment, when the Cascades, especially in Washington, developed a distinct rain shadow. In eastern Oregon, the vegetation, notes Greg Retallack, closely resembled today's grassy live oak woodland and savannah on the western slopes of the Sierra Nevada in northern California. Other trees included hickory, maple, water chestnut, and mountain mahogany. The animals looked more like what you would find in an African- or Asian-themed wildlife park. Gomphotheres—short burley elephants with both upper and lower tusk—would have rummaged through live-oak-laden forests. And a close cousin, the mastodon *Zygolophodon*, browsed on riparian leaves. Multiple varieties of Shetland pony-sized horses still had three toes, though the middle toe was well on its way to becoming a hoof. Camels—looking more like llamas than true ships of the desert—shared the landscape with rhinos and peccaries. Predators included the last of the nimravids (animals that resembled saber-tooth cats but lacked retractable claws and had a different auditory canal) as well as wolf- and coyote-sized ancestral dogs. Beavers built dams across streams where catfish swam. Based on the clays and structures of Late Miocene soils, Retallack has estimated that annual rainfall approached 30 inches—much more moist than today's 18 inches. Seasons were mild. But that would soon change.

Goat Rocks, between Mount Adams and Mount Rainier, is the glacially eroded remains
of a Pliocene volcano that would have rivaled these two High Cascade volcanoes in size.

CHAPTER 10 The Big Chill *Shifting Climates*

If purgatory could be a geologic time period, it would be the Pliocene. This was a time undistinguished by major eruptions, extinctions, accretions, or meteoritic bombardment. But the Earth moved toward the coming ice age. Between 5.3 and 2.6 million years ago the planet cooled, pausing, 3.2 to 2.8 million years ago, for a time of CO_2-driven warming, often considered an analogy for today's anthropogenic climate change.

In the Pliocene, the Isthmus of Panama swung into place. Ocean currents in the Indian and Pacific oceans changed. And the world waited breathlessly for the Ice Age to enter, stage right.

This is not to say that nothing happened in the Northwest. The Basin and Range continued its expansion. Lakes appeared in the deepening valleys. The clockwise rotation of the Coast Range, Cascades, Klamaths, and Willamette Valley continued. Horses committed to their present single-hoofed form. The most noteworthy Pliocene action

occurred in the Cascades, building the foundations of present peaks. These included Hannegan caldera, sequestered in the North Cascades, just north of Mount Baker, which produced a mighty eruption about 3.7 million years ago. Goat Rocks volcano, northeast of Mount Adams produced rhyolites, tuffs, and some basalts, 3.2 million to about 1 million years ago.

In Oregon and southcentral Washington, Pliocene eruptions occurred east of today's main High Cascades axis. Underwood Volcano, which forms layered cliffs along the Columbia at White Salmon, Washington, dates to the Early Pliocene (and Late Miocene), about 6 to 4 million years ago. Farther to the east, about 4 million years ago, the Simcoe Volcanic Field near Goldendale produced hot basalt lavas that rose rapidly and explosively from the mantle. Lorena Butte, a privately owned cinder cone that serves as a rock quarry, yields some of the rarest of terrestrial rocks—fresh chunks of the Earth's mantle, called mantle xenoliths (from the Greek: *xeno*,

stranger; and *lithos*, stone). Mantle xenoliths are mostly found in volcanoes with deep soures (mantle plume) for its lavas. Hawai'i—specifically, Kilauea—is littered with them. Their presence in the Simcoe volcanics is mysterious. The Simcoe vents seem an ordinary set of volcanoes, compared with Hawai'i. But we do know, based on the compositions of the xenoliths and research by Bill Leeman of Rice University, that the Earth's mantle beneath Goldendale has been steeped in potassium- and sodium-enriched fluids rising from the subduction zone 20 miles or so below the Goldendale City Hall. Whether this contributes to the presence of xenoliths is another question.

Farther south, the Central Cascades also produced errant, off-axis eruptions. They include Snow Peak, a Pliocene volcanic center 25 miles east of Albany, Oregon. Radiometric dating indicates that it is approximately 3 million years old. Several Pliocene basaltic andesite shields form a ridge just west of the

The Simcoe volcanics include Lorena Butte, just outside Goldendale, Washington, where basaltic lavas include small chunks of the Earth's upper mantle, like this pristine bit of peridotite, with very fresh olivine.

High Cascades, southwest of Mount Jefferson. The best-known volcano of this group is Iron Mountain, which towers above US Hwy 20 near Tombstone Pass. The cliffs of Iron Mountain and nearby Browder Ridge expose bedded cinders intruded by dikes and sills. Similar rocks make up Three Pyramids, a Pliocene andesitic center about 20 miles west of Mount Jefferson.

Evidence for Pliocene eruptions in the Cascades proper is found in 4- to 5-million-year-old lava flows west of the Cascade crest, including Frissell Point and Bunchgrass Mountain, above the McKenzie River near McKenzie Bridge. In southern Oregon, Mount Thielsen and Diamond Peak sport bases of Pliocene-age lavas. The remnants of Pliocene eruptions occur along the Klamath River Gorge. These basalts have the chemical signature of rift lavas, rather than true subduction zone/Cascade volcanics. They may be Basin and Range lavas, or they were generated along a rift related to the Cascade graben. Similar Pliocene volcanic rocks (basalts, andesites, and pyroclastic deposits) about 4 to 2.5 million years in age continue south between Mount McLoughlin and Mount Shasta.

To the north, the Washington Cascades were also active, building volcanoes that now are the foundations of modern peaks. Goat Rocks, a low-slung, craggy wilderness between Mount Adams and Mount Hood, began erupting in the Pliocene about 3.2 million years ago. At its most majestic, the Goat Rocks volcano may have rivaled Hood, with a summit towering 10,000 feet in altitude. Alas, its eruptions waned by about 730,000 years ago, and Ice Age glaciers made short work of the isolated peak. Today Goat Rocks is truly a shell of a volcano, with huge glacial moraines leaving telltale tracks of the icy thieves that stripped away its glory.

As the Northwest's climate became cooler and wetter, rivers grew and the erosive power of water increased. The Pliocene is known as a time of major canyon cutting. Hells Canyon, deepest gorge in North America (yes, deeper than the Grand Canyon if you measure from the summit of He Devil Peak to the adjacent Snake River more than 7,400 feet below) was pioneered when Pine Creek, an unprepossessing small stream, captured a river that ran north and emptied into Lake Idaho. Lake Idaho, a vast, shallow body of water, that covered most of the modern Snake River Plain, provided habitat for 8-foot-long saber-toothed salmon. Today, Lake Idaho's legacy is a mile-thick formation of siltstones and sandstones. These soft sedimentary rocks are well exposed at the Pillars of Rome, northwest of Rome, Oregon, and related formations closer to Ontario, Oregon.

Middle Pliocene Climate: Back to the Past?
Much of the Pliocene was warmer than present climates. As such, many scientists consider the Pliocene, especially the Middle Pliocene (3.2 to 2.8 Ma), to be a model for where the current global climate is headed.

The warmth of the mid-Pliocene is attributed partly to atmospheric compositions and partly to oceanic circulation, including the closure of the Isthmus of Panama and the end

The Pillars of Rome in southeastern Oregon are Pliocene stream and lakebed sediments of the Rome Formation. They contain gravels, sands, and volcanic tuff that represent beach, stream, and deeper water deposits.

of a permanent global El Niño. Atmospheric carbon dioxide, based upon isotopes, plant stomata, and other indices, hovered between 360 and 440 ppm, and global mean temperatures were about 5 degrees F higher than temperatures today. As in today's climate-change models, polar regions were substantially warmer then (up to 15 degrees F, with greater warming in the winter, according to NASA simulations). Further warming at high latitudes comes from the increased levels of atmospheric water vapor (a greenhouse gas) that results from the warm, ice-free ocean conditions, whereas equatorial temperatures were essentially unchanged. In the Arctic, Pliocene forests dominated where tundra exists today. Pliocene Arctic soils were wetter than the present day, fed by increased rainfall originating over the warmer Arctic Ocean.

Today, carbon dioxide is above 400 ppm, and will reach the Pliocene's 440 ppm long before the year 2100, based upon current predictions. Globally, Pliocene continental distribution, oceanic circulation, and ecosystems were similar to modern situations. So the climates and ecosystems of the Pliocene provide a potent guide to our fairly immediate future.

Based upon Pliocene ecosystem distribution, as CO2 levels rise toward 440 ppm we can anticipate a global increase in temperature, higher precipitation in the Northwest and in many now-arid regions, a spread of tropical savannahs and woodland in Africa and Australia as deserts shrink, and a northward shift in boreal forests, according to a 2009 paper by Saltzman and others in the Royal Society's Philosophical Transactions.

Although our climates may revert to Pliocene conditions, our extinction-ravaged Anthropocene wildlife diversity will not equal the flagrantly odd diversity of the Pliocene. The fauna of the Pliocene would have inspired a modern double take. Closure of the Panamanian land bridge about 3 million years ago allowed the first-ever interchange of mammals between North and South America—an episode known as the Great American Biotic Interchange. Camels, tapirs, gomphotheres (elephant family), horses, canids, and cats migrated to South America. In exchange, we got giant, flightless carnivorous birds, ground sloths, and opossums. (Perhaps not the best of deals.) Fauna found at McKay Reservoir near Pendleton, Oregon, include ducks, quail, sandpiper, and two species of camels. Findings at other localities across the Northwest include two different species of horse, antelope, a small grasslands rhino (*Teleoceras*), a mastodon, beaver, marmot, a small fox-sized dog or coyote (*Canis condoni*), a larger canid more like a dire wolf with jaws more adapted to bone-crushing, and an ocelot-sized cat.

Pluvial Lakes in the Desert

Beginning in the Pliocene and continuing into the Pleistocene (from at least 5 million years ago to about 8,000 years ago) areas of southeastern Oregon, which are now stark deserts, were lush wetlands and extensive lakes. These water bodies, known as pluvial lakes, which develop during geologically extensive periods of heavy rainfall, appeared throughout the Basin and Range. Lake Bonneville, whose diminished relics are best known as the Great Salt Lake and Bonneville Salt Flats, is perhaps the most familiar.

Oregon's pluvial lakes supported robust ecosystems. The lakes hosted a flourishing population of warm-water catfish, suckers, and Tui chub. They watered mammals large and small, from Columbian mammoths and cameloids to antelope and horses. They were framed by trees, and invaded by cattails and extensive marshlands. And late in their history (Late Pleistocene to Early Holocene, 15,000 to 2,000 years ago), these pluvial lakes were a favorite human habitation, including perhaps for refugees driven south from the Columbia Basin by devastating floods. Evidence of human settlements surrounding the lakes includes 15,000-year-old remains in Paisley Cave, 10,000 year-old sandals found in Fort Rocks Cave, and more sophisticated villages of stone such as 2,300-year-old Carlon Village. The immediate sources of water that created this high-standing lake level were elevated precipitation during the last glacial maximum and, eventually, glacial meltwater.

Fort Rock Lake was the largest PNW pluvial lake. At its maximum extent, lake waters covered 1,240 square miles, with a maximum depth of 320 feet. Fort Rock lay near the northern shore. Wind-driven waves breached the original Fort Rock tuff cone, eroding wave-cut notches into the side. Famously, Luther Cressman found sandals woven from sagebrush bark and dating to about 10,000 years in Fort Rock Cave, on the lake's northern shore.

Fort Rock, in Oregon's High Desert, displays a wave-cut bench that records the level of pluvial Fort Rock Lake. Pluvial lakes in the Great Basin region persisted from the Pliocene into the late Pleistocene. Fort Rock itself is a Pleistocene feature with an estimated age of less than 100,000 years.

The second largest of the Pacific Northwest pluvial lakes was Lake Chewaucan. It included the present Abert Lake (Oregon's largest saline lake) and Summer Lake. Lake Chewaucan extended across about 450 square miles, from the present Valley Falls north to Alkali Lake, and west past today's town of Summer Lake. Its maximum depth, achieved about 12,000 years ago, was about 375 feet.

During the Pliocene, a large lake and wetland occupied today's Baker Valley in northeast Oregon. It was evidently a popular hangout 4 to 5 million years ago. A bluff excavated behind the Always Welcome Inn has provided the remains of many plants and animals to Jay Van Tassel and his students at Eastern Oregon University. The fossils include diatoms, reeds, tree branches and roots, and leaves. More importantly, the waters supported clams, snails, ostracods (tiny crustaceans), minnows, sunfish, and (probably) chub. This large, generally shallow lake extended 15 miles from Baker City to Keating. Its shores supported hares, weasels, rails, ducks, gophers, shrews, voles, two species of beaver, and even a llama-like camel. Fauna at other Pliocene locations in the Northwest, especially the renowned, 3.5 million-year-old Hagerman Fossil Beds in Central Idaho, includes horses (*Equus simplicidens*), bone-crushing dogs, mastodons, and saber-toothed cats. It is likely that these creatures also roamed the uplands behind the Always Welcome Inn.

Altogether, pluvial lakes and other waters covered more than 4,000 square miles of

The Ringold Formation forms the White Bluffs along the Columbia River's Hanford Reach. These Pliocene sediments include the fossils of muskellunge and sturgeon.

southeast Oregon's desert, and northeast Oregon's valleys. Especially in the southeast, they persisted through the Pleistocene, providing food and support for some of the earliest human habitations known in the new world, 14,000 to 15,000 years ago. Today, they are saline lakes rimmed by dusty playas, with ancient shorelines etched far above our heads.

Far to the north, along the great bend of the Columbia River's Hanford Reach, stark white cliffs overlook the nation's first nuclear reactor. The White Bluffs are composed of the Ringold Formation (4.5 to 3 Ma)—sands and clays deposited by the ancestral Columbia and tributaries, some fine-gained shallow lake deposits, as well as floodplain and deltaic deposits. The formation is exposed in or underlies much of the Pasco Basin, suggesting that the riparian/floodplain system that deposited the sediments was extensive.

This lake-strewn river system appealed to plants and animals of the time. The White Bluffs include the remains of fish such as sturgeon, muskellunge, lake sucker, bullhead catfish, and sunfish. The only salmon is the 6- to 8-foot-long "saber-toothed" salmon (*Oncorhynchus rastrosus*, also classified as *Smilodonichthys*) that occurs in sediments at the base of the White Bluffs. Gerald Smith and colleagues suspect that a thermal barrier may have excluded later salmonids from traveling this far up the Columbia system. Other animals entombed in the Ringold's White Bluffs include rabbits, shrews, muskrats, cats, bear, horses, camels, wild boars, and deer.

The cataclysmic eruption of Mount St. Helens in 1980 focused our attention on the potential activity of many High Cascade volcanoes, including their potential for explosive eruptions, lahars, and summit collapse.

CHAPTER 11   The Young and the Restless   *High Cascade Volcanoes*

The High Cascades are the quintessential Northwest landscape. Rugged, snow-capped, swaddled in lush forests, steeped in Pacific maritime weather, these iconic peaks define our regional psyche. Picturesque Mount Hood. Stately Mount Rainier. Bombastic Mount St. Helens. Azure and sacred Crater Lake. And the towering mystery of Shasta. The volcanic range stretches about 700 miles, from Garibaldi Peak in British Columbia to Lassen in northern California. Altogether, more than 3,000 separate volcanoes, large and small, make up this volcanic range. They are young volcanoes: even the most iconic peaks are less than 800,000 years old.

The early history of the High Cascades includes the calm eruptions of basaltic shield volcanoes, now largely buried beneath the higher peaks, and powerful blasts of ignimbrites that raced eastward into the Deschutes and Columbia Basins. In central Oregon, faulting pulled the range apart beginning about 7 million years ago. These rifts unleashed a wide variety of ash, tuffs, explosive eruptions, and quiet basalts that filled the Deschutes Basin to the east. The faults dropped the older basaltic shield volcanoes into a faulted depression—the High Cascades Graben. This structure has long since been filled in by subsequent eruptions, but it is evident in geophysical surveys, appearing as a gravity anomaly, traced by occasional seismic shudders. Today, the eastern flanks of a few of the low, basalt-rich early High Cascade volcanoes are exposed on Green Ridge, northwest of Sisters, and the basalt flows of the 7-million-year-old shield volcanoes make fleeting appearances in the canyon of the Deschutes River at Cove Palisades State Park.

The Pleistocene marks the rise of the Cascade peaks familiar to us today. The High Cascades are mostly a range of composite volcanoes. They are built of lava flows that alternate with ash, debris flows, and tuff to build steep-sided, charismatic cones. Their sculpted profiles reflect their battle with Ice Age glaciers that wore them away at the same time that eruptions sought to build them higher. Through time, the High Cascades have become more explosive as their magmatic systems evolved from fluid basaltic lavas to viscous dacite. Today, the lavas most likely to erupt are the most difficult to disgorge. One look at Mount St. Helens' sheared dacite dome is a reminder that this lava does not erupt with ease.

The oldest of the High Cascades include smaller peaks, well preserved by virtue of location on the arid east side of the range, that may never have supported glaciers. They include Black Butte (1.2 Ma), west of Sisters, Oregon, and portions of the Simcoe and Indian Heaven Volcanic Fields east of Mount Adams, which date to the Pliocene. Farther north, these older eruptive centers include Kulshan caldera (1.2 Ma), a rhyolitic caldera marooned on ridges between Mount Baker and Mount Shuksan.

During the Pleistocene, numerous High Cascade peaks grew to large size, only to lapse into dormancy and extinction before the glaciers melted away. Consequently, the volcanoes could not replace or repair the scoured, erosional damage of moving ice. Glaciers had their way with Oregon's Mount Thielsen, Mount Washington, and Three Fingered Jack, deeply eroding their cones and revealing the mountains' inner structures. These once rivaled Mount Jefferson in size; today their skeletal remnants provide backdrops for scenic hikes and hazardous climbs. Only the peaks that erupted sufficient volume to replace the rock worn away by glaciers stand as iconic highpoints on the horizon today.

The iconic volcanoes of today's High Cascades began the series of eruptions that built the present cones and landscape beginning only 730,000 years ago. To put this in a more human-friendly time scale, the lavas that would become Mount Hood began building the present-day mountain about the same time that *Homo heidelbergensis* appeared in Europe, and before both Neanderthals and *Homo sapiens*. It was the middle of the Pleistocene.

The High Cascade's oldest eruptions date to the Middle Pleistocene, about 1 million years ago. Mount Baker seems an exception to this, with lavas about 1.3 million years in age. On Rainier, the oldest known rocks of the modern cone are andesites along Rampart Ridge about 500,000 years in age. Hood's oldest eruptions of the modern cone date to about 600,000 years—andesites exposed near Cooper Spur. An older version of the mountain, Sandy Glacier volcano, is exposed on Hood's western flank, and dates to about 1.2 million years. At Mount Mazama (Crater Lake) the oldest remaining portion of the main cone is Phantom Ship, some 700,000 years old. Shasta and Jefferson are among the youngest of the large volcanic cones, with an age of about 300,000 for its oldest lavas. Mount Lassen, essentially a huge dacite lava dome, dates to only 28,000 years, although Brokeoff Mountain, part of the original and larger peak scoured away by glaciation, dates to 300,000 years.

Today, the Washington Cascades sport five large, widely spaced andesitic composite volcanoes. In contrast, Oregon's Cascade Range includes six slightly smaller volcanoes that tend to be more basaltic. In California, the andesitic behemoths of Shasta and Lassen dominate the skyline. Basaltic volcanism of the southernmost Cascades seems diverted to the Modoc Plateau and the huge shield volcano of Medicine Lake.

Recipe for Cascade Lavas

High Cascade lavas are typical products of a subduction zone. Andesites and pyroclastic flows predominate. Basalts and rhyolites are less abundant. The following complicated process generates most eruptions:

1. The Pacific seafloor moves down the subduction zone, carrying with it oceanic sediment, altered basalts, and a lot of water.
2. As the down-going slab reaches depths of 35 to 50 miles, pressure and heat drive water from the oceanic rocks.
3. Water rises from the subduction zone into the overlying mantle where it acts as a flux, allowing the mantle to melt at lower temperatures than normal.
4. The melted mantle is a magma with the composition of basalt.
5. The fluid basalt rises toward the surface to erupt as a basalt. More commonly, it encounters and incorporates silica-rich components from the melted slab.
6. If the resulting fluid can make it to the surface, an eruption ensues.

There are infinite possibilities for the end product, from a fluid mantle melt of basalt that rises so quickly that it never mixes with anything, to the highly mixed andesite and dacite lavas produced by Mount Lassen's 1917 eruptions. Some—probably most—lavas never make it to the surface. South Sister provided considerable excitement in 2002, but the rising plume of basaltic magma, tracked by geochemistry, seismicity, and a subtle uplift, never erupted, despite being cheered on by volcanologists around the globe.

The dominant rock of the Cascades is a volcanic rock called andesite. Usually a medium-gray color, this very ordinary-looking rock is the de rigueur ingredient of composite

Andesite is the most common volcanic rock in the Cascades. It is usually gray in color, may display large, often irregular columnar joints, and commonly forms thin, horizontal "platy" joints like these in an outcrop on the flanks of Mount Baker.

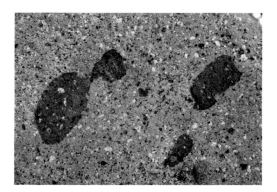

Explosive eruptions like that of the 1915 events at Mount Lassen, or Crater Lake, 7,700 years ago, are often triggered by the injection of a hot basalt magma into the magma chamber beneath the volcano. Here, the light-colored dacite erupted from Lassen in 1914 to 1915 includes evidence of this trigger as globs of much darker basalt.

volcanoes that rise above subduction zones worldwide. Its name comes from its prevalence in the Andes, but it dominates many Cascades peaks. Andesite is often mistaken for the sedimentary rock shale because it separates into very thin, horizontal "platy" joints, as well as columns that are often much wider than the more regular columnar jointing of basalts.

Many Cascade volcanoes have a history of early basalt eruptions, followed by thicker, less mobile but more voluminous andesites, and finally, gooey, silica-rich dacites. This sequence reflects several volcanic facts of life. The first is that basalt is a very fluid lava that may reach the surface quickly. Secondly, the production of andesite generally requires melting and assimilation of other components. Third, andesite is more viscous than basalt, and reaches the surface more slowly. And lastly, dacite is even more viscous and requires greater amounts of melting of the crust, and a lot of water to keep it mobile.

These same factors contribute to the explosive late stages of many composite volcanoes. Viscous dacite and rhyolite lavas act as plugs or corks that solidify before the magma chamber is emptied. Gasses—mostly steam and sulfur, with some carbon dioxide, build up pressure. Add a pulse of hot basaltic magma and the system overheats, blowing the dacite "cork" out of the vent—and often the entire mountainside. This scenario triggered Lassen's 1917 blast as well as Mount Mazama's cataclysmic eruption 7,700 years ago.

Beacon Rock and the Boring Volcanics

Beacon Rock, dated at 57,000 years, is the youngest volcanic feature in the Columbia River Gorge, and the youngest of the Pleistocene-age Boring Volcanic Field, which occupy the Portland Basin and extend eastward into the Cascades. Beacon Rock rises abruptly 848 feet above the Columbia River.

At an age of only 52,000 years, Beacon Rock is the youngest of the Boring Volcanic Field, small volcanoes that pepper the Portland area.

Some 50,000 years ago it would have been a cinder cone on the banks of the Ice Age Columbia. But by 10,000 years ago, Ice Age Floods had eroded all the cinders, leaving only the scoured basaltic core (neck) of the cone, and some red scoria near the top to attest to its volcanic past. If you climb the long, winding stairway to the top, you'll find telltale oxidized scoria in trail-side outcrops near the summit, which only the frothing fury of eruption can craft.

More than eighty Boring volcanoes pepper the Cascade foothills and Portland Basin. Kelly Butte, Rocky Butte, Mount Scott, Mount Sylvania, Sylvan Hill, Highland Butte, Larch Mountain, and Powell Butte are some of the Portland peaks. In Washington, Green Mountain, Underwood Hill, Prune Hill, and Battleground State Park are among the easiest to find. Their compositions are diverse, ranging from andesites at Larch Mountain (1.4 Ma) to basaltic andesites at Rocky Butte (285,000 years old) to a primitive, alkali-rich basalt scoria at Mount Tabor (205,000 years old).

The Boring volcanoes are the westernmost potentially active volcanic field in the United States. Although not officially part of the High Cascades, they are cousins, also generated by arc-related processes. The present vents are considered extinct. However, the overall Boring Volcanic Field is not, as noted by Russ Evarts, Richard Conrey, and other geologists who have studied the area. New volcanoes could appear in the future.

Mount Baker

Mount Baker (10,778 feet) is among the Cascade's most active volcanoes and is the second-most-heavily glaciated peak, with about 0.4 cubic miles of ice in the glaciers and snowfields of its summit region, as of 2000. This composite volcano is located about 30 miles east of Bellingham. Unlike most other Cascade volcanoes, which have erupted atop the ruins of older volcanic systems, Mount Baker rises directly above the much older (Paleozoic and Mesozoic) accreted terranes of the North Cascades.

Mount Baker and nearby related vents have been almost continually active for the past 1.3 million years. One long-extinct vent area, known as Chowder Ridge, apparently produced andesites, dacites, and rhyodacites from 1.3 million years ago until about 100,000 years ago. Glaciation during the last, Fraser glacial advance, erased much of the Chowder Ridge vent area, leaving only sills, dikes, and well-polished remnant flows. Baker's earliest eruptions are also recorded in the Kulshan caldera just to its northeast, where ash-flow

Mount Baker, at 10,778 feet in elevation, straddles the North Cascades, and has a high potential for future eruptions.

tuffs and rhyolites erupted between 1.3 and 1 million years ago. The second and succeeding major eruptive complex dates to 1.1 million to 600,000 years in age, and produced mostly andesites and dacites. The third generation of eruptive materials dates to 500,000 to 200,000 years ago. These andesites are exposed on Ptarmigan and Lasiocarpa Ridges. The Black Buttes stratocone has produced basaltic to dacitic lavas between 500,000 to 200,000 years ago. Mount Baker's most recent eruptions have come from the stratocone and contemporaneous peripheral volcanoes 100,000 years to the present.

Mount Baker was once more massive. Conservative estimates suggest that Mount Baker has produced almost 40 cubic miles of lava flows and associated eruptive products. Andesite and rhyodacite each make up nearly half of the volume erupted, whereas basalt and dacite represent less than 10 percent. About 6,600 years ago, ash-rich eruptions shook the peak. In 1975, increased fumarolic activity in the Sherman Crater area caused concern that an eruption might be imminent. The activity has increased slightly since 1975, suggesting that Mount Baker may awaken sometime in the future.

Glacier Peak

Glacier Peak, 10,525 feet, has neither the massive stature of Rainier nor the accessibility of Mount Baker. But this unobtrusive volcano, only 70 miles from Seattle and tucked into the Glacier Peak Wilderness, ranks among the most active peaks in the High Cascades. During the past 3,000 years, it has produced at least five major eruptions. Most have erupted viscous dacite lava, which may plug a vent temporarily, allowing pressure to build, with an explosive eruption the end result. At least two of these produced more than a half-cubic mile of ash, which likely rose to heights of about 20 miles into the atmosphere, according to James Gardner and colleagues at the University of Rhode Island. The most recent explosive eruption occurred around the year 1700. The ash layers from these and earlier eruptions provide geo-chronological marker beds throughout the Northwest. Lahars followed the Skagit River to Puget Sound as recently as 1,800 years ago. Previous mudflows clogged the Stillaguamish River as recently as 13,000 years ago, providing an unstable and slide-prone substrate, and also underlie much of the city of Arlington. The U.S. Geological Survey considers Glacier Peak a "very high threat volcano with inadequate monitoring."

Mount Rainier

Mount Rainier is the highest (14,410 feet), third-most voluminous (64 cubic miles), and, unquestionably, the most dangerous volcano in the Cascades. It threatens the Seattle area with eruptions and devastating lahars (mud and debris flows).

Rainier's history of destructive behavior includes the collapse of its summit and

Emmons Glacier carves into the northeastern slopes of Mount Rainier. The mountain has rebuilt it 14,410-foot summit after a summit collapse event 5,600 years ago sent avalanches of debris into Puget Sound.

generation of hot debris flows that reached Puget Sound during an eruption only 5,600 years ago. This collapse formed a large crater much like that at Mount St. Helens in 1980. Ensuing eruptions rebuilt Rainier's summit, filling the large collapse crater. Today, Rainier looks stately, but beneath its elegant edifice, ongoing alteration and decay of seemingly solid rock threaten the repeat of the same devastating events.

Rainier, like other composite volcanoes, has a long and complex history. The earliest known deposits of Rainier are about 2.9 million to 840,000 years old. These rocks are mostly ash and debris flows found west of Rainier. They are known as the Lily Formation, and may represent a precursor to the true "Rainier," rather that the first products of Rainier itself.

Most of Mount Rainier's cone was built by a remarkably uniform series of multiple andesite flows, some more than 100 feet thick. These lava flows are interspersed with ash, debris flows, and other volcaniclastic rocks, making Rainier a true composite volcano. These eruptions included a major ash eruption sometime between 70,000 and 30,000 years ago that spread several hundred miles to the south and southeast. Rainier produced significant eruptions of ash 2,350 years ago, and also produced the summit, Columbia Cone, less than 2,200 years ago, based upon ash found beneath the cone. Geologist Dwight Crandell has identified more than fifty-five

lahars and debris flows of Holocene age from Mount Rainier. Rainier last erupted lava about 1,000 years ago.

Present thermal activity includes occasional steam explosions on upper slopes where many steam vents are located. Multiple sulfurous steam vents at areas above 10,700 feet, including one at the summit, suggest that today, Rainier's hydrothermal system remains alive and threatens future action.

The Osceola mudflow, 5,500 years ago, covered at least 100 square miles, and varies in thickness from 20 to 200 feet. The mudflow extended into a former arm of Puget Sound—"former," because the mudflow filled the area and changed the coastline, pushing the shoreline 20 miles westward and creating 180 square miles of new land.

The much more recent Electron Mudflow, named for the small community of Electron which is built on the flow, occurred about 550 years ago. This lahar, which evidently began as a failure of part of the altered rock on the western side of Mount Rainier, has not been correlated with any eruptive activity at Rainier and may have occurred without a linked eruption.

Both of these mudflows indicate that Rainier holds serious threats to the towns clustered at its feet. Enumclaw, Auburn, Buckley, and the Port of Tacoma rest on the Osceola lahar. Orting and Puyallup are built on the Electron Mudflow. It is best to keep an eye on the restive giant to the east.

Mount St. Helens

Mount St. Helens was just another pretty face in the Pacific Northwest, until everything changed on May 18, 1980. This young composite volcano finally earned its stripes as a major Cascade eruptive center, and helped volcanologists craft a new understanding of the Cascades.

Mount St. Helens' oldest known lavas and ash date to only 50,000 years ago, making it the youngest major volcano in the Cascades. The rocks that formed the now-vanished symmetrical cone and summit were between 2,500 and 100 years old—astoundingly young.

The first eruptions of Mount St. Helens, known as the "Ape Canyon" stage for the area where the rocks are best exposed, produced silica-rich lavas—andesite and dacite. The next stage of Mount St. Helen's construction began only about 13,000 years ago. Known as the Swift Creek stage, the eruptions produced large volumes of ash and pyroclastic flows. By 10,500 years ago, the eruptions dwindled. Mount St. Helens lay dormant between 10,500 years ago and 4,000 years ago.

The construction of the modern volcanic carapace (at least the pre-1980 visage) is known as the Spirit Lake stage of Mount St. Helens' life. Six separate phases of eruptions built the modern cone. These eruptions were more basaltic—they contained more iron-rich lavas, and on the whole were less explosive than previous eruptions. True basalts, including those of the Ape Cave flow on Mount St. Helen's south side, erupted about 2,000

While areas more distant from Mount St. Helens have recovered from the 1980 eruptions, the immediate flanks of the volcano remain barren.

years ago. The largest volume appeared on the south; both other basalts erupted from the mountain's base on the north. These fluid pahoehoe lavas poured from vents at the base, while the summit continued to build dacite lava domes. Ape Cave is a mile-long lava tube—one of many lava tubes in these flows.

Three other periods of eruptive activity mark St. Helens' history prior to its reawakening in 1980. One, about 1,200 years ago, produced the only lateral blast known at St. Helens prior to the 1980 event.

On March 20, 1980, it awoke again with a magnitude 4.1 earthquake. Then, on May 18, at 8:32 a.m., a magnitude 5.1 quake about 1 mile beneath the volcano triggered an enormous landslide that slid rapidly off the north side of the mountain. Within a few minutes the summit of the mountain was reduced from its 9,677-foot elevation to the present 8,363 feet, as 3.7 billion cubic yards of rock slid away. The slide unleashed a powerful lateral blast of ash and gas. It is likely, notes volcanologist Don Swanson, that the release of confining pressure allowed volcanic gas to jet into the air, and also permitted superheated groundwater in the volcano to flash to steam. The blast exploded from the mountain at a velocity of more than 300 miles per hour, and was supersonic close to the mountains, at temperatures in excess of 660 degrees F. The blast carried 250 million cubic yards of the volcanic edifice, dispersing the material and torrid gas over 230 square miles, and as far away as 17 miles from the summit. The eruptive column reached a height of 80,000 feet within 15 minutes of the triggering earthquake.

Ultimately, the ash covered 22,000 square miles, falling in significant amounts as far away as Walla Walla, Washington. Lahars rushed down the Toutle River, destroying homes and eviscerating bridges along the banks, carrying logs and log trucks, ultimately clogging the Columbia with 20 feet of mud, rocks, debris, logs, and log trucks. The May 18 eruption included pyroclastic flows that swept from the abbreviated summit and engorged crater at up to 80 miles per hour, at temperatures of 1,300 degrees F. Altogether, these ash flows accumulated to depths of up to 120 feet. Much of the north side of the mountain today rests as a lumpy, hummocky terrane. It is analogous to a similar landscape on the north side of Mount Shasta. Until the Mount St. Helens eruption, the genesis of the Shasta landscape was not understood. Now it is seen as an enormous debris avalanche likely unleashed in a similar collapse and eruption.

Mount St. Helens remains active, including the eruption of a new dacite dome, beginning in 2005. Its future activity may be confined to dome building, to more explosive eruptions, or a lapse into volcanic slumber and senility.

Mount Adams and the Simcoe Volcanic Field

Mount Adams (12,281 feet) has a total volume of more than 50 cubic miles, second only to Mount Shasta. Like Shasta and most Cascade volcanoes, Adams is composed primarily of andesite. Located about 30 miles east of Mount St. Helens, the summit area of Mount Adams hosts twelve rapidly shrinking glaciers.

Mount Adams is considered an active volcano. This impressive volume includes more than sixty small volcanoes and lava flows of the Mount Adams Volcanic Field south

Mount Adams, a massive peak, rises to 12,281 feet and looms above the surrounding landscape. Sulfur was mined from areas near its summit in the early twentieth century. Much of the summit area is highly altered and prone to collapse.

and east of Mount Adams. This field, which includes the Indian Heaven area, began erupting 940,000 years ago, with a major pulse of eruptions 120,000 to 100,000 years ago. The basalt-rich volcanic field spans about 480 square miles.

Mount Adams proper was built in three main eruptive episodes. The earliest eruptions that would build the massive peak occurred about 520,000 years ago. A second phase of major eruptions started 450,000 years ago, also producing andesite. Along Cascade Creek, as many as forty andesitic lava flows are exposed, spanning the time from 457,000 to about 320,000 years ago. The third major eruptive phase is dated at 30,000 years ago. Summit eruptions during this phase encountered glacial ice, producing scoria, spatter cones, and angular glassy blocks. Adams has remained continuously active on a smaller scale throughout its history. The andesites and dacites that form the south summit are about 15,000 years old. Those on the true summit may be slightly younger, at 13,000 years. The summit area of Mount Adams contains steam vents and fumaroles that vent hydrogen sulfide (H2S). The most recent documented eruptions are four small ash eruptions that occurred about 1,000 years ago.

Adam's unstable summit area has produced large lahars and debris flows within the past 10,000 years. These include a lahar of about 90 million cubic yards generated 6,000 years ago, which inundated the Trout Lake lowland and continued down the valley of the White

Salmon River at least as far as Husum, more than 35 miles from Mount Adams. Today, these deposits are 3 to 65 feet thick. This light-colored deposit is visible as a sediment layer in the banks of the White Salmon River. Large rocks transported by the debris flow have been stranded in low-lying fields along the river. The most recent large event occurred in 1921 when about 5 million cubic yards of crumbly, altered rock fell from the summit area and rushed about 4 miles down Salt Creek Valley.

The 200-square-mile Simcoe Volcanic Field extends across most of the Simcoe Mountains, and lies 30 miles east of Mount Adams, generally near Goldendale, Washington. Its eruptions range in age from about 5 million years to less than 500,000 years, and consist of rhyolite, dacite, and basalt. The oldest lavas are rhyolites found in the easternmost part of the Simcoe Volcanic Field. Signal Peak is a dacite dome. Previously associated with Mount Adams, the Simcoe Volcanic Field taps a deep mantle source for many of its basaltic lavas. Examples include Lorena Butte, where pristine mantle xenoliths have been brought to the surface by the eruptions of this (privately owned) cinder cone.

Adams and its nearby smaller siblings rank among the best candidates for destructive future activity in the High Cascades. Steam vents pepper the mountain's summit area, which was mined for sulfur in the early 1900s. Like Mount St. Helens, Adams has likely entered its dotage as an active Cascade peak. Future eruptions—or attempts at eruptions—will likely involve viscous dacite or rhyolites, likely making any volcanic event a dramatic and destructive one.

Mount Hood

Mount Hood is Oregon's highest peak at 11,239 feet. Its oldest lava flows may be the 700,000-year-old andesites exposed at Cooper Spur on the east flank of the mountain. Andesites and dacites dominate Hood. Mudflows, domes and fumaroles, and pyroclastic flows and debris also contribute to the mountain's bulk, making Portland's backyard peak a classical composite volcano. Its overall composition is 70 percent lavas and about 30 percent ash, mudflows, and other unconsolidated volcanic materials. Four eruptions during the past 15,000 years have spilled pyroclastic block-and-ash flows and lahars into the four river systems (Sandy, Salmon, Hood, and White) that drain the mountain and its twelve glaciers.

Mount Hood, like Mount Rainier, sports a summit of crumbling, altered rock that could collapse, producing far-reaching lahars and debris fields. This not-unlikely scenario is a significant hazard to communities near the volcano. And like Rainier, Hood offers proof of past debacles. Deposits that are rich in altered Mount Hood andesite underlie terraces in the lower Hood River valley (20 miles north of the volcano) and are present on both sides of the Columbia River near the town of Hood River, approximately 30 miles north of the volcano. This ancient Pleistocene lahar presumably began as a debris avalanche that incorporated large masses of preexisting rock on the flank of the volcano, inundated the Hood River Valley, and temporarily filled the Columbia River to a depth of almost 100 feet. A single radiocarbon date from wood in the lahar is greater than 38,000 years. Old lahars are also exposed 25 miles west of Mount Hood in the Sandy River drainage. Weathered zones that are more than 20 feet thick indicate that the deposits west of Mount Hood are many tens of thousands of years old.

Hood's most recent eruptive phase, the Old Maid eruptive period, began with emplacement of the Crater Rock hornblende dacite dome high on the cone. Numerous lahars probably fed by avalanches from the dome and accompanying snowmelt entered the Sandy, Zigzag, Salmon, and White Rivers; a pyroclastic flow traveled from the Crater Rock area at least 4 miles along the White River, burying and toppling trees as it went. One lahar extends more than 35 miles down the White River, partly filling the upper White River canyon near Timberline Lodge. A terrace made of a lahar overlain by reworked eruptive debris is more than 40 feet thick on the lower Sandy River, 30 miles from the mountain. The Sandy River lahars buried the cedar forest at Old Maid flat, setting the stage for the flat, lodgepole pine–landscape present today. Lost Creek and other streams have exhumed stumps and other remnants of this buried forest. Tree ring–based dating

Mount Hood, 11,289 feet, is an active volcano that erupted about 1793, producing significant lahars, and Crater Rock—a dacite dome—just a few years before Lewis and Clark made their appearance.

by Patrick Pringle of Central Washington University indicates that the Sandy River lahar occurred in the mid-1790s, the pyroclastic flow in the upper White River about 1800, and the lahar that traveled 50 miles down the White River between 1800 and 1810.

Crater Rock remains a locus of steam and sulfur emissions today. Present thermal activity is in fumarole fields near Crater Rock, at the apex of a semicircular zone of fumaroles and hydrothermally altered heated ground. In summer 1987, maximum ground temperatures were near 185 degrees F and maximum fumarole temperatures were about 200 degrees F, slightly above the boiling point of water at 9,300 feet. Many of the fumaroles are actively precipitating crystalline sulfur. Comparison of modern and historical photographs shows that the amount of perpetually snow-free ground surrounding the fumarole fields has been increasing since last century. Mount Hood seems ready for action.

Mount Jefferson, 10,495 feet, is considered a very low risk for future eruptions. Its last eruption was about 27,000 years ago.

Mount Jefferson

Oregon's second highest peak (10,495 feet), Mount Jefferson hasn't erupted for more than 27,000 years. Still, by geologic standards, where the second hand ticks in millennia, this composite volcano is merely dormant, rather than extinct. Its eruptions began at least 1 million years ago, according to Rick Conrey and other geologists who mapped and studied the volcano extensively. The high-rising Central Cascades cone dominates the Madras skyline.

It is built mostly of andesite flows and dacite domes.

The Mount Jefferson Volcanic Field covers about 60 square miles. It includes many small domes and basaltic shield volcanoes. Because crustal extension here allowed many avenues for magma to reach the surface, lavas spread across a wide area. This contrasts with Mount Hood, where only one major path to the surface developed, building a larger and more singular structure.

Mount Jefferson consists of more than 160 separate eruptive units. The main cone consists of andesites and dacites domes and short flows. The ramparts around the main cone include more basalts and basaltic cinder cones. An older (2 to 3 Ma) silica-rich eruptive center is present about a mile northeast of Jefferson's summit. Another, about 4 to 5

million years in age, lies to the west according to Rick Conrey.

The silica-rich lavas, including dacites, are generated at depths of 15 to 18 miles by melting of subducted and metamorphosed seafloor basalts. Andesite is generated by a similar process—melting of highly pressured and metamorphosed subducted basalts—at greater depths of 20 miles beneath the volcano. The presence of granodiorite clasts in some pyroclastic rocks suggests that a small and extinct magma chamber underlies the volcano, according to Conrey.

Thick crust and high heat flow are necessary to produce the lavas of Mount Jefferson. The crust beneath Mount Jefferson is about 25 miles thick, though whether it is composed of accreted terranes or Siletzia, or a combination, is uncertain.

Three Sisters and Collier Cone

The Three Sisters volcanic complex in central Oregon consists of five large volcanoes: North, Middle, and South Sisters; Broken Top; and the Husband. Numerous cinder cones, domes and small lava flows are scattered across the more than 100-square-mile field. North Sister is the oldest, longest-lived, and most mafic of the Three Sisters, with mostly dark, basaltic lavas 574,000 to about 170,000 years in age; Middle Sister (37,000 to 14,000 years) and South Sister (178,000 to a little less than 2,000 years) have erupted basalt to rhyolite magmas. South Sister is considered active, along with vents near North Sister.

A major fault system—the High Cascades Graben—underlies Three Sisters, as well as other basaltic Cascade volcanoes, including Mount Washington and Belknap Crater. This system of small (mostly 2- to 5-mile-long) faults extends from Three Sisters north to Mount Adams, down-dropping the crest of the central High Cascades as much as 1,000 feet. It has persisted for at least the last 8 million years, and widened the central Cascade crust as much as 1.5 miles. Today, it is still widening in response to overall extension of the Pacific Northwest's crust. Recent estimates of extension rates by USGS scientist Ray Wells indicate that each year Sisters and Salem are about 1 mm farther apart. The fault system allows basaltic lavas to reach the surface more frequently and in more locations than other segments of the Cascade arc.

North Sister is an anomaly among major Cascade volcanoes. It is one of the few mafic volcanoes in the chain. Its dark, rubbly peak is composed exclusively of basalt, basaltic cinders, and a basalt-like—but slightly more silica-rich—rock called "basaltic andesite." The summit area is laced with dikes—more solid and resistant to erosion than the surrounding cinders. North Sister belongs, according to Mariek Schmidt and Anita Grunder of Oregon State University, to a distinct class of volcanoes—along with other examples such as Stromboli Volcano—which are relatively monotonous and long-lived mafic (iron-magnesium-rich, mostly basaltic) arc volcanoes. The volcano was constructed in four eruptive stages, at 400,000 years, 164,000 years, and 90,000 years, and more recent cinder cones—15,000 to as little as 1,600 years ago—that include Collier Cone.

Little Brother, a small shield volcano just west of North Sister, erupted during the last Pleistocene glacial advance and was severely eroded by glacial ice. Its lavas interfinger with and bank against North Sister's flank. Although basaltic, Little Brother's lavas came from a completely different source than those of North Sister, and are similar to the lavas erupted by Mount Washington.

After construction of North Sister, basaltic lavas erupted along an elongate fissure—where Matthieu Lakes lie now. The fissure produced lava flows and spatter cones, including the hill just east of North Matthieu Lake, about 15,000 to 10,000 years ago. These lavas "erupted through and were banked against"

North Sister and Middle Sister are the most mafic (iron-rich) of the Three Sisters group of volcanoes. The Middle Sister, mostly andesite, last erupted about 14,000 years ago. North Sister unleashed basalt until 170,000 years ago, although young cinder cones on its flanks produced most of the basalt lavas at McKenzie Pass less than 5,000 years ago.

glacial ice that occupied the ridge and basin along the fissure area, according to Schmidt and Grunder.

Nearby cinder cones, including Collier Cone, Four-in-One Cone, and Yapoah Cone, are quite young. At 1,600 years, Collier Cone, with lavas ranging from basalt to andesite, is among the youngest volcanic centers in Oregon. Their lavas, like those of Little Brother, are more similar to Mount Washington than North Sister.

Middle Sister is younger and smaller than North Sister, with dates on its largely andesitic eruptions ranging from 37,000 to 14,000 years, with most activity from 25,000 to about 17,000 years ago according to U. S. Geological Survey researcher Wes Hildreth. Its lavas have a broad range of compositions, from andesite and dacite to basalt. Several dacite domes and vents remain prominent, including The Black Hump (aka Step Sister) on the ridge between North and Middle Sister, which has been dated at 27,000 to 18,000 years. Obsidian Cliffs, about 5 miles to the east of the peak, may have been the first eruption of lavas related to Middle Sister, 30,000 years ago.

South Sister is composed of much more silica-rich lavas than its siblings to the north.

It is also the youngest, and most likely to blossom into eruptions. South Sister's oldest lavas are young, even by High Cascade standards, according to research by Hildreth and Fizzar, and include a dacite on its southeast flank dated at 178,000 years. Andesites on its northeast flank have been dated at about 93,000 years. Between 50,000 and 30,000 years, South Sister produced mostly rhyolites—including those of Devils Hill and Kaleetan Butte as well as unnamed features to the north and east. Andesite eruptions occurred from 38,000 to 32,000 years ago, including the emplacement of an intrusion near the mountain's summit.

The latest eruptions on South Sister occurred in two episodes about 2,000 years ago. They formed the gray, glassy dacite domes known as Rock Mesa on the southwest flank of South Sister, and Devils Chain—a series of sixteen domes and stubby, glassy dacite flows on the southeast flank of the mountain. Initial explosive eruptions produced small pyroclastic flows and tephra fallout from several aligned vents low on the south flank, according to William Scott of the U.S. Geological Survey. Then, lava emerged from two vent areas, forming a large lava flow, Rock Mesa, and several small lava domes. Decades to a few centuries later, a similar eruptive sequence

South Sister is the youngest and most potentially active of the Three Sisters. It last erupted about 1,200 years ago, producing Rock Mesa and The Devils Chain—a series of rhyolite domes.

occurred along a linear series of vents, the Devils Chain, which extends from just north of Sparks Lake to high on the southeast flank of South Sister. About the same time, a shorter zone of vents developed on the mountain's north flank near Carver Lake. Geologists think a silica-rich magma chamber remains beneath South Sister, and that future, possibly explosive, eruptions are a reasonable bet, or,

Mount Bachelor, a favorite of Central Oregon skiers, is a basaltic cinder cone built upon a shield volcano base. Wetlands abound in the high, flat marshes of the area.

as Wes Hildreth notes, "magmatism beneath the Three Sisters region remains potentially as vigorous as ever."

In 2003, South Sister threatened an eruption. About 3 miles west of the volcano, an accumulation of a modest volume of magma, at a depth of about 3 to 4 miles, raised the ground surface several inches over a broad area about 10 miles in diameter. Geologists promptly dubbed it "The Bulge." Magmatic gas, including trace amounts of sulfur dioxide and helium, lent credence to the idea that an eruption might be imminent. In May 2003 a series of earthquakes suggested that magma, gas, and/or steam were rising beneath the bulge. However, tremors ceased after a day of (promising) activity, and today the bulge area appears quiet.

Mount Bachelor

Mount Bachelor (9,068 feet) is directly south and compositionally the opposite of South Sister. Rather than silica-rich rhyolite, Mount Bachelor is a cone of iron-rich basaltic rock. The U.S. Geological Survey classifies its broad, basaltic base as a shield volcano. Known as Kwohl Butte, the broad base of basaltic lavas is well exposed to the south and the west of Bachelor. The summit cone that we ski and hike on—Mount Bachelor proper—is largely cinders and blocks of basaltic andesite, with slightly more silica than plain basalt. With a total volume of about 6 cubic miles, Bachelor is the northernmost and largest of a series of nearly fifty vents known as the Mount Bachelor Chain. These cinder cones,

shield volcanoes, and basalt flows stretch for about 15 miles almost north-south, and are about 18,000 to 8,000 years old. The oldest is Sheridan Mountain. The youngest eruption from the Mount Bachelor Chain may be the lava flow that created Sparks Lake by blocking the Deschutes River, and is traversed by the Ray Atkeson Trail.

Newberry Volcano

First, and most important, Newberry is not truly part of the Cascades. It is a shield volcano that is part of the High Desert volcanics, and as such, owes its genesis to the Brothers Fault Zone rather than to the Cascadia Subduction Zone. However, as magmatism slowly works its way west to intersect with the Cascades beneath South Sister, Newberry is, in some ways, a 7,984-foot-tall, High Cascade kissing cousin.

Newberry's 1,200-square-mile footprint— an area nearly the size of Rhode Island— ranks it as the most extensive volcano in the continental United States. (Geologists who need to describe the sizes of things have long been grateful for the New England states.) Newberry's lavas extend 75 miles north-south and at least 35 miles east-west. It is a low-slung shield volcano bristling with cinder cones. Its summit exhibits the ragged remnants of 500,000-year-old rhyolite torn apart by explosive eruptions 75,000 years ago, leaving two jewel-blue lakes and crenulated obsidian flows to fill the gaping summit depression.

Newberry volcano is the largest basaltic shield volcano in the United States. Its oldest eruption occurred more than 740,000 years ago. Its most recent eruption produced the 1,200-year-old Big Obsidian Flow, shown here.

Pyroclastic flows and tuffs in Teepee Draw, dated at 500,000 years, are conventionally accepted as the oldest rocks at Newberry. But rhyolite tuffs on Newberry's eastern flank may be significantly older according to USGS geologist Julie Donnelly-Nolan. These flesh pink pyroclastic rocks are reversely magnetized, indicating they are older than 730,000 years. Basalts recovered from drill cores that seem related to Newberry have been dated as about 1.8 million years old.

Newberry has a bombastic history. At least three calderas have been blasted out of the volcano's center. The earliest corresponds with the tuffs of Teepee Draw, which indicates that the initial eruptions of Newberry were sometime before 500,000 years ago. A second collapse, recorded in sediments and tuffs found in drill cores, dates to 400,000 to 200,000 years ago. The present caldera was crafted by an explosive eruption 75,000 years ago. Between these destructive eruptions, Newberry continued to build its low-lying profile with basalt eruptions, as well as construction of more than 400 cinder cones, especially on its northern flanks.

Newberry has been busy recently. Less than 7,000 years ago it produced the Interlake Obsidian flow and Central Pumice Cone that divide Paulina Lake from East Lake. Faulting along the Northwest Rift Zone began soon after, providing entrée for fluid basalts that built Lava Butte, Lava River Cave, and Lava Cast Forest. Newberry's most recent eruption, only 1,300 years ago (AD 700) produced the Big Obsidian Flow. This volcano is considered an eruptive threat today, as well as a target for geothermal energy generation.

Crater Lake/Mount Mazama

Crater Lake is possibly the most famous composite caldera in the world—and certainly the most photographed and celebrated. Its deep caldera was formed by collapse during the catastrophic eruption of a huge volcano, Mount Mazama, about 7,700 years ago. The cataclysmic eruption produced approximately 12 cubic miles of magma, ash, and rock fragments. Prior to its climactic eruption, Mount Mazama's summit likely exceeded 11,000 feet. Today its highest point, Mount Scott, struggles to reach its 8,000-foot height.

Mount Mazama was a multi-spired peak, built of several cones perhaps similar to modern Mount Shasta. Rocks of the Phantom Ship—now a lonely rock outcrop that juts from Crater Lake's southern end—represent the earliest known eruptions. These andesites are approximately 400,000 years old. Subsequently, at least six major centers erupted combinations of mafic andesite, andesite, or dacite before approximately

Crater Lake was formed by the explosive eruption of Mount Mazama about 7,700 years ago. Klamath Indian accounts of the eruption indicate that it lasted for three days. Llao Rock on the northern shore is a massive dacite flow that erupted just before the final cataclysm.

On the southern flanks of Mount Mazama, The Pinnacles in Sand Creek Canyon showcase the reason for the explosive eruption. The earliest ash is light-colored dacite. The top part is darker, and is composed of basaltic ash. This dark material represents hot basalt magma that was injected into the mountain's large dacite magma chamber, triggering the cataclysmic eruption. Similar features occur along Annie Creek.

75,000 years ago. Between 75,000 and 50,000 years (during the Pleistocene), andesitic and dacitic lavas erupted from five or more vents. During much of the Pleistocene, basaltic eruptions built cinder cones on the flanks of Mount Mazama. Basaltic magma from one of these vents, Forgotten Crater, erupted

sometime between approximately 30,000 and 22,000 years ago. Dacite domes appeared near Mount Mazama's summit between 50,000 and 22,000 years ago. The dacitic Palisade flow on the northeast wall is approximately 25,000 years old. These lavas probably represent eruptions in the early stages of development of the climatic magma chamber.

Approximately 8 to 12 cubic miles of Mount Mazama was blown away or collapsed into the crater during the climactic eruptions 7,700 years ago. These eruptions emptied virtually all the magma in the molten reservoirs beneath the mountain. Evidence for this paroxysmal purge comes from the spectacular compositional zonation in late deposits of

Mazama's eruptions at Annie Creek and The Pinnacles. They display the light-colored rhyodacite from the top of the magma chamber, and grade into more and more iron-rich dark material that represent a hot basaltic magma that, injected into an already roiled and active magma chamber beneath the mountain, likely triggered the cataclysm.

The ash and pumice of the Mazama eruption was carried predominantly to the northeast; only sparse, thin patches of re-deposited pumice are present southwest of the caldera rim. Ash-fall must have occurred over at least several continuous days, based upon the thickness, and the time required for dispersal. The Mazama ash is 20 inches at Newberry Volcano, 68 miles to the northeast, and exists as a thin layer less than 1/2 inch in southwestern Saskatchewan, 745 miles from its source. The ash is commonly found in northeastern Oregon and Washington as small light-colored deposits that mark the locations of old stream channels.

Geologists have not yet declared Mount Mazama dead. Post-collapse eruptive activity, including Wizard Island and the submerged Merriam Cone, has kept the main crater of the once-mighty volcano alive. And to geologists, this means that Mazama's plumbing is still intact—and likely aimed at the lake bottom. In volcanic life cycles, 7,700 years is the equivalent of a refreshing nap.

Mount Shasta

Mount Shasta (14,162 feet) is the High Cascades most massive volcano. It rises just west of the town of Weed, California, about 40 miles south of the Oregon border. It is a landmark, visible from Klamath Falls, from the slopes of Mount Ashland, and from higher points in the Kalmiopsis Wilderness. It is the second highest of the Cascades, after Mount Rainier. Part of Shasta's bulk can be attributed to a secondary volcano, Shastina, which rises to more than 11,000 feet on Shasta's west flank, making Shastina alone the fourth-tallest peak in the Cascades.

The smooth-sided, picturesque cone of Shasta that we see today is a volcano built only during the past 100,000 years. It rises atop the ruins of a 600,000-year-old massive volcano that was destroyed by "sector collapse"—the catastrophic collapse of the entire northwest face, resulting in a massive avalanche—sometime between 360,000 and 300,000 years ago. This massive slide and debris flow traveled more than 30 miles north of the summit and covers about 250 square miles. The odd, lumpy topography of Shasta Valley was recognized as a huge, blast-generated landslide deposit in 1984 after geologists noted that when the north side of Mount St. Helens slid away, the event produced a similar, though much smaller, hummocky deposit. In fact, the avalanche that destroyed the ancestral Mount Shasta was twenty times larger than the deposits at Mount. St. Helens.

The oldest part of the present mountain is exposed at Sargent's Cone on Shasta's south flank. Composed of andesites, ash, and lahars, Sargent's Cone represents a rapid series of eruptions sometime between 200,000 and 100,000 years ago. This volcano grew quickly, and also erupted for a short period of time—perhaps only a few thousand years according to the U.S. Geological Survey. Erosion, including glacial ice, removed much of the original Sargent's Ridge volcano by about 20,000 years ago, although small eruptions of pasty rhyolitic lavas, ash, cinders, and mudflows continued.

The main edifice of Mount Shasta, known as the Misery Hill Cone, began its eruptions between 50,000 and 30,000 years ago. These eruptions, initially of andesite and ash, built a composite cone that buried much of the older Sargent's Ridge cone. The summit of Misery Hill was capped by eruption of dacite domes, including Hotlum Cone, which today forms the summit of Shasta.

Hotlum Cone's activity includes lava flows and ash that has erupted within the past 6,000 years. Its irregular eruptives have occurred on average every 600 years, with the last recognized eruption in 1786, when a rising column of ash was described by the explorer La Perouse. This observation is corroborated by carbon-14 dates on trees buried by ash east of the volcano at about 200 years.

Shastina's eruptions began less than 10,000 years ago, building the peak to its 11,000-foot elevation rapidly. Carbon-14 dates suggest that Shastina may have been built in only a few centuries' time. This newly minted volcano

Mount Shasta, 14,162 feet, is a massive composite volcano. It lost much of its summit to a summit collapse event about 300,000 years ago and has built its present elegant profile since then. Shastina, the smaller cone on its northwest flank, rises to 11,000 feet, and has been built in the last 10,000 years.

was shaken by explosive eruptions of pyroclastic flows and the eruption of more domes at its summit.

Black Butte, a complex of four large dacite domes that rises to about 6,000 feet on the west flank of Shastina and is easily visible from Interstate 5, is dated at only 9,500 years. Debris and ash-flow tuffs from this eruption extend 7 miles to the base of Mount Shasta.

Part of Shasta's looming threat comes from the broad apron of domes that extend 7 to 10 miles north and east, suggesting that magmas are dispersed and capable of eruptions across the region. This is compounded by the presence of abundant basalts and basaltic andesite dikes and flows, which indicate that access to hot lavas may be one of Shasta's fortes.

Mount Shasta has continued to erupt at least once every 600 to 800 years for the past 10,000 years. Its most recent eruption probably was in 1786. Proof of this eruption, recorded from sea by the explorer La Perouse, is somewhat ambiguous, but his description could only have referred to Mount Shasta. A small crater-like depression in the summit dome, containing several small groups of fumaroles and an acidic hot spring, might have formed during that eruption; lithic ash preserved on the slopes of the volcano and widely to the east yields charcoal dates of about 200 years.

Mount Lassen

Mount Lassen is the southernmost major volcano of the Cascades, and one of the most recently active. Its last major eruptions in 1914 to 1917 produced the full gamut of volcanic products, including a small lava flow and pyroclastic flows (hot ash flows), as well as mudflows and debris flows.

Like many Cascade volcanoes, Mount Lassen marks an area that has been active for a long time—specifically, about 3 million years. Mount Lassen itself began its eruptions about 600,000 years ago. Initially, between 600,000 and 400,000 years ago, Lassen built a huge single andesitic stratovolcano, known as Brokeoff Volcano. As part of its last series of eruptions about 400,000 years ago, Brokeoff produced several dacite domes—harbingers of things to come. The final eruptions of Brokeoff jettisoned more than 10 cubic miles of silica-rich rhyolite ash and pumice—explosive eruptions of tuff and ash flows. Based upon the size and distribution of the remnant peaks, we can estimate that at its maximum size, Brokeoff Volcano was approximately 10,500 feet high, had a basal diameter of approximately 7 miles, and a volume of about 20 cubic miles. Potassium-argon ages of lavas from Brokeoff Volcano range from 0.59 to 0.39 million years ago. Thus, Brokeoff was active for approximately 200,000 years.

Mount Lassen produced two more episodes of eruptions. Between 250,000 and 200,000 years ago, Lassen adopted a different eruptive style, which included lavas that contained more silica, and hence were more viscous and explosive. These lavas built a dozen large domes, including Bumpass Mountain, and produced about two cubic miles of ponderous, gray, and often glassy lava flows. Rather than erupting andesite, as was common in the old Brokeoff Volcano, the Lassen dome field has produced mostly dacite—a glassy, viscous lava. Dacite lavas of the Lassen Volcanic Field frequently contain inclusions of andesite—fragments of the roots of the older, extinct, and eroded Brokeoff Volcano.

The modern eruptions of the Lassen Volcanic Field, including building the volcanic dome we know as Mount Lassen, began about 100,000 years ago and continue today. Lassen Peak itself—about 25,000 years old—is a product of the most recent eruptive sequence, as are Eagle Peak (55,000 years) and Chaos Crags (1,050 years). Glacially sculpted Lassen Peak, which rises about 2,000 feet above the surrounding landscape, was likely built during 1 or 2 years of eruptions, and is one of the largest volcanic domes on the planet.

Until Mount St. Helens reawakened in 1980, Lassen was the most active volcanic center in the continental United States. Lassen Peak's most recent eruptions occurred from 1914 to about 1921, beginning on May 30, 1914, with a small steam eruption near the summit of the peak. More than 150 explosions of

Mount Lassen (center) is actually a huge dacite dome, part of a cluster of domes that erupted beginning about 100,000 years ago.

various sizes occurred during the following year. By mid-May 1915, the eruption changed in character. Lava flowed from the summit crater about 100 yards over the west and probably over the east crater walls. Ensuing eruptions produced mudflows, pyroclastic flows, and columns of ash that rose more than 4 miles into the atmosphere and deposited ash some 200 miles to the east. The lavas of this eruption were rhyodacite and dacite in composition. They include darker splotches of basalt within the dacite. This indicates that the explosive eruption was likely triggered by an injection of hot basalt into the magma chamber below. A similar event—injection of hot basalt into a dacite chamber—also triggered the cataclysmic eruption of Mount Mazama about 7,700 years ago.

Lassen has remained dormant, but sullen. Steam rising from Bumpass Hell and the malodorous Sulfur Works attests to the fitful nature of volcanic slumber.

Medicine Lake Volcano
(Lava Beds National Monument)

Medicine Lake Volcano is the sort of mountain that you can easily miss. It is only half as high as Shasta (7,762 feet), and its gentle slopes are peppered with trees and cinder cones. This unprepossessing shield volcano is actually the largest of any Cascade volcano by volume, covering about 250 square miles and erupting an estimated 130 cubic miles of basalt and related lavas. This dwarfs Rainier (60 cubic miles erupted in the past million years) and even makes mighty Shasta (85 cubic miles)

Lava Beds National Monument includes Captain Jack's Stronghold, where Modoc warriors, outnumbered ten to one, held out against U.S. Army forces for months in 1873. Medicine Lake Volcano (distant left) is a vast shield volcano at the southern end of Lava Beds National Monument. Like Newberry, it is not produced by subduction-related processes, and although close in proximity, is not considered part of the Cascades. Mount Shasta is the white peak visible in the distance.

seem puny. Medicine Lake's last major eruption, about 950 years ago, produced Glass Mountain—a 0.8-cubic-mile obsidian dome on the east slopes of the mountain.

Medicine Lake's lavas are, as expected, dominantly mafic, with about 75 percent basalts or basaltic andesites. Rhyolite and rhyodacite cover less than 10 percent of its area, according to Julie Donnelly-Nolan. However, at depth, the volcano is very different, with a core of 50 to 60 percent silica-rich lava flows and domes. This silica-rich core is about 450,000 to 300,00 years old. Beneath these flows is a granitic intrusion (dated at 319,000 years) that may be more than 3 miles in diameter. Hence, the early history of Medicine Lake Volcano was that of a voluminous silica-rich eruptive center, followed by even more voluminous basalts, beginning about 300,000 years ago. Basaltic injection also provides high heat flow at Medicine Lake Volcano, making

it an attractive target for geothermal energy today.

Within the last 85,000 years, most eruptions from Medicine Lake have been dacites, including Glass Mountain and Little Glass Mountain. Among the exceptions to this pattern is the Giant Crater Lava Field and Basalt of Valentine Cave about 11,000 years ago, and multiple basalt flows about 5,000 years in age on the east and northeast flanks of the volcano. These flows include the rugged terrane of Captain Jack's Stronghold. The basalt flows on the mountain's north and east side were emplaced largely by flowing through lava tubes as inflated flows. Some of these lava tube systems extend 15 to 20 miles from the vent system. Lava Beds National Monument contains the largest concentration of lava tubes known in North America.

During the Pleistocene (Ice Age), continental glaciers advanced into northern Washington. Near Withrow, Washington, oversized erratic boulders spread among wheat fields mark their terminal moraine.

Global Temperatures

Millions
of years   2500   PROTEROZOIC   541   PALEOZOIC   252   MESOZOIC   66.0   CENOZOIC

Camb | Ordo | Silur | Dev | Miss | Penn | Perm | Trias | Juras | Creta | Paleo | Eoce | Olig | Mioc | Plioc | Pleist | Holo | Anth

CHAPTER 12   Glaciers and Floods   *The Ice Age*

The Pleistocene, more commonly known as the Ice Age, marks a time of repeated glacial advances and retreats. It began 2.59 million years ago, when climate shifted decisively toward colder global temperatures, and lasted until 11,700 years ago. The term "Pleistocene" was coined in the 1820s by Charles Lyell, and comes from the Greek roots *pleistos* (most) and *kainos* (recent), when Lyell understood it as the time we live in today.

In the Northwest, the Pleistocene is divided into multiple stages—or times when ice advanced and times when ice retreated. Although the precise dates of advance and retreat vary from place to place, owing to elevation, latitude, and other factors, the generally acknowledged glacial advances that affected the Cordilleran Ice Sheet and regional alpine glaciers coincide with stages named in the Puget Sound area: the Orting (2.4 to 1.6 Ma), Salmon Springs (1.06 Ma), Possession glaciation (80,000 years), and Fraser (22,000 to 10,000). These correspond roughly to the Laurentian ice sheet stages of Nebraskan, Kansan, Illionian, and Wisconsin. Dating older glaciations is challenging because organic material trapped in moraines is beyond the dates of about 50,000 years where carbon-14 can be used, and later glaciations often erase or overprint evidence of previous glacial advance and deposition.

Pleistocene Alpine Glaciers in the Northwest

Alpine glaciers are crenulated streams of ice that groan their way from the high flanks of mountains to the greener valleys below. Or at least they used to. Today, alpine glaciers, worldwide and in the Northwest, are fast disappearing under the onslaught of climate change. But at least four times in the past 2.5 million years, glaciers sculpted mountains with impunity.

Some of the most classical and easily accessible Pleistocene alpine glacial features in the Northwest are found at northeast Oregon's Wallowa Lake and southeast Oregon's Steens Mountain. Wallowa Lake and its moraines track glacial advances and retreats over as much as 300,000 years. The inner, youngest moraines tower almost 900 feet above the lake surface. No other alpine moraine system in North America is as well exposed, as accessible, or offers as complete a record. Research has not yet dated the oldest of the moraines—a series of ridges that extend eastward, known as the complex East Moraine, that can best be observed from the Prairie Creek Cemetery—but scientists have estimated their age at 200,000 to 300,000 years.

Only 17,000 years ago, a glacier occupied Wallowa Lake. Ice towered several hundred feet above today's moraines. Geologist Joseph Liccardi used beryllium isotopes generated by cosmic rays to determine that large granite erratics atop the east moraine had been in place for 17,200 years, and those on the west moraine had been stationary for an average of 16,700 years. So, on average, the last time a very large glacier could have carried these big

Alpine glaciers sculpted the mountains of eastern Oregon. Repeated advances and retreats left a complex series of lateral and terminal moraines flanking Wallowa Lake.

boulders and deposited them atop the lateral moraines along today's lake was 17,000 years ago. Where the East and West Fork of the Wallowa River meet, glacial ice was about 1,500 feet thick. At its terminus, ice towered to 1,200 feet in thickness, at least 400 feet above Old Chief Joseph's gravesite high on the terminal moraine. Smaller, recessional moraines mapped by Liccardi in the Wallowas indicate that by 10,000 years ago, the last glaciers had retreated to the Wallowa Mountains' Lakes Basin.

At Steens Mountain's summit you can visit another kind of glacial landscape—erosional.

Here, a cap of glacial ice covered the mountain. Alpine glaciers polished much of the exposed summit rock. Walk across the landscape west of the summit and above the great U-shaped canyons, and you'll find striated rock surfaces polished to a glossy sheen. The polishing agent? Glacial ice full of sand and sharp-edged stones. From Steens' broad summit area, classical U-shaped canyons radiate to the north, west, and south. Although many classical U-shaped glacial valleys, including those of the Wallowas, are now obscured by forest, the landforms at Steens Mountain are open and evocative of the raw might of glacial ice.

In the Olympic Mountains, the last advances of the Pleistocene are part of the Fraser ice advance and are subdivided into three distinct advances or "stades": the Evans Creek Stade (21,000 to 19,000 years), the Vashon Stade (15,000 to 13,600), and the rather wimpy Sumas Stade, at 11,500, that coincided with a global cold surge when the North Atlantic Conveyor (Gulf Stream) was disrupted by Ice Age Floods that poured from the St. Lawrence River. The Hoh and Elwha Rivers, among many others, inhabit these oversized, U-shaped valleys today.

Continental Ice Sheets in the Northwest

The great continental ice sheets that covered northern North America were divided into two separate glacial systems—the Laurentian glacier east of the Rockies and the Cordilleran glacier on the west side. The Cordilleran continental ice sheet advanced into northern Washington and covered virtually all of the Puget Sound area during the glacial maximums of the Pleistocene. Glacial striations and abandoned erratics are evident in the Seattle and Tacoma areas. Terminal and recessional moraines and glacial tills mark their tracks in eastern Washington.

The record of Early and Middle Pleistocene glaciations is obscure because glaciers tend to obliterate their previous tracks with new advances. Hence the best records we have for the extent and distribution of glacial ice was during the Fraser stage of the Pleistocene—80,000 to about 11,000 years ago—with glacial maxima about 16,000 to 14,000 years ago (though the precise age for the maximums vary with altitude and latitude).

During the last advance, five separate lobes of the Cordilleran continental glacier edged into Washington. The Puget lobe scoured the Seattle area and the Juan De Fuca lobe moved westward around the Olympic Mountains, joining with the alpine glaciers that moved down from the peaks. Three lobes were present east of the Cascades: the Okanogan lobe, Columbia River lobe, and Purcell Trench lobe. The extensive Okanogan lobe covered the Methow Valley and Okanogan uplands, and also invaded the northeasternmost part of the North Cascades. Ultimately, this lobe dammed the Columbia River, forming Glacial Lake Columbia, and ultimately contributing to the great Ice Age Floods of Glacial Lake Missoula. The Columbia River lobe also dammed the river to the east. The Purcell Trench lobe, easternmost of the well-defined lobes, moved south into the Pend'Oreille area, damming the Clark Fork River with a 2,000-foot-high wall of ice and leading to the multiple Missoula Floods.

The long, winding hills that form terminal moraines allow geologists to map the farthest extent of these glacial lobes. The terminal moraine of the Okanogan lobe appears on the Waterville Plateau, just west of the Grande Coulee. Known as the Withrow Moraine for the small community of Withrow nearby, it marks the farthest advance of the Okanogan lobe with a row of gravelly mounds. The nearby Jameson Lake Drumlin Field is an extensive landscape of streamlined gravel hills that mark the direction that the ice moved.

West of the Cascades, the Fraser/Vashon advances of ice reached a maximum at about 14,0000 to 14,500 years. This date is much younger than likely maximum dates for higher elevations regions in the interior of Oregon, where alpine glaciation at Wallowa Lake may have reached a maximum at about 16,500 years ago, with glaciers in retreat by 14,000.

Ice Age Floods in the Pacific Northwest

In 1926, when J. Harlen Bretz first recognized evidence of Pleistocene floods that swept across eastern Washington, he envisioned a single event at the end of the Ice Age. Today geologists have defined at least forty, and possibly more than a hundred separate torrents that swept across the Columbia basin when an ice dam repeatedly formed and failed, again and again releasing hundreds of cubic miles of water stored in Glacial Lake Missoula. And recently, geologists including Richard Waitt, Bruce Bjornstad, Kevin Pogue, and Scott Burns have defined even more ancient floods that raged across the Columbia Basin as long ago as 2 million years.

Evidence for ancient floods include gravels found beneath more modern flood sediments in roadcuts west of Walla Walla, Washington, and coulees at Connell, Washington, and west of Connell at Vernita Grade and Washtucna. Farther afield, older floods left evidence at Yakima Bluffs and in roadcuts south of The Dalles, Oregon. Estimated ages of these older

Scratches and polish on bedrock in the Wallowa Mountains were produced by glacial ice during the Pleistocene.

floods range from 200,000 to 400,000 years old, based on radiometric dates and the presence of thick layers of caliche, to older than 780,000 years, based on the reversed magnetic polarity of the deposits. Even older deposits have normal polarity, which would place them as older than 1.77 million years. Bruce Bjornstad has suggested that some features that we attribute to recent (15,000-year-old) floods may be much, much older. In one giant flood bar, up to 300 feet thick, the age of the deposits indicate that the bar had grown to half its present height by 780,000 years ago. The sources, frequency, number, and intensity of the older flood events are as yet unknown.

The most recent Ice Age Floods that ripped through the landscape began about 15,000 years ago and concluded by 13,000 years ago, leaving a sodden and sculpted terrain that humans would soon claim. In fact, humans may well have witnessed these Late Pleistocene events. The oldest human presence in North America dates to about 15,000 years—at Paisley Cave in southcentral Oregon. Native tribes very likely lived along the salmon-rich Columbia River. Traditional Yakama stories include the value of high places such as Rattlesnake Mountain (3,527 feet) as refuges from big waters. In fact, Rattlesnake Mountain's Yakama name, Laliik, means Land above the Waters.

The Ice Age Floods from Glacial Lake Missoula and Glacial Lake Columbia are detailed in numerous books and videos. Only a brief summary is included here. Many other

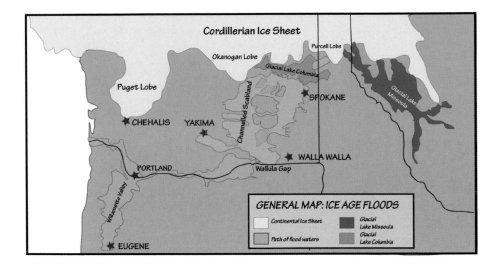

Ice Age Floods from both Glacial Lake Missoula and Glacial Lake Columbia burst across the Pleistocene landscape.

glacial outburst events have been documented in the Northwest during the Late Pleistocene, including a flood from the overflow of Lake Bonneville (Great Salt Lake) into the Snake River that raged through Hells Canyon, and outbursts and overflows of water from Puget Sound that carved an oversized channel along the Grays River to the sea. Worldwide, outburst floods were common and often rivaled those of the Northwest in scope and volume. One such flood, an outburst from Glacial Lake Agassiz (now the Great Lakes) poured so much water down the St. Lawrence River and into the Atlantic that it shut off the North Atlantic Conveyor (Gulf Stream) about 11,500 years ago, causing a return to much colder, glacial climatic conditions known as the Younger Dryas.

The Missoula Floods developed when the southernmost of the Purcell lobe of the Cordilleran continental ice sheet blocked the drainage of the Clark Fork River, backing water up to form Glacial Lake Missoula behind a 2,000-foot-high dam made of glacial ice. At its greatest extent, this doomed, ephemeral lake drowned the site of today's Missoula, Montana, beneath 950 feet of water and covered 3,000 square miles. In the biggest floods, almost 500 cubic miles of icy, turgid water would reformat the Columbia Basin's landscape.

Each time the ice dam failed, the Purcell Lobe would rebuild, blocking the Clark Fork River once more and eventually unleashing yet another flood. In each flood event, water surged across Rathdrum Prairie northeast of Spokane, leaving giant ripple marks. The flood continued across central Washington, stripping off soils, eroding canyons into bedrock, and generally transforming a fertile, flat prairie landscape into the bizarre, barren-channeled scabland. Noteworthy features sculpted by the fast-moving waters include the Grande Coulee, Dry Falls, Drumheller channels, Potholes, and Palouse Falls. Estimated water velocities exceeded 65 mph.

The Grande Coulee, almost 1,000 feet deep, up to 3 miles wide, and more than 30 miles long, served as a conduit for Ice Age glacial outburst floods mostly from Glacial Lake Columbia, a body of water that was essentially a larger version of the lake impounded behind Grand Coulee Dam today. The lake formed when the Okanogan lobe blocked the Columbia's channel. The water was released when the ice dam failed. The floods from Glacial Lake Columbia coursed through Moses Coulee and Grand Coulee, producing steep-sided and unforgettable canyons.

Dry Falls is one of the most renowned and spectacular features of the Ice Age Floods. At 3 miles wide and 400 feet high, with a double falls (and a triple falls in the biggest floods), it is hard to miss. But although an imposing topography today, during maximum flood events Dry Falls was only a ripple. The water surface was almost 300 feet above the present top of the falls at velocities estimated at 65 mph.

Dry Falls is a classic, if oversized, example of a waterfall that has retreated upstream, leaving a canyon behind. Before Ice Age Floods began, there was a small falls about 17 miles

downstream near what is now Soap Lake. The falls' retreat upstream was accomplished during multiple large flood events, including floods that occurred in the Early Pleistocene as well as the multiple, better documented Late Pleistocene events. Richard Waitt suggests that as many as a hundred floods may have been large enough to significantly erode and drive the retreat of Dry Falls from Soap Lake to its present position—a rate of retreat that would require only about a quarter mile of retreat during each major flood.

Like Dry Falls and the canyon it crafted to the south, the Grande Coulee also represents a canyon etched by the retreat of a waterfall. Grand Coulee's falls have now vanished, and all that remains is rubble. However, Steamboat Rock represents a fragment of the original falls' headwall. Richard Waitt and colleagues calculate the Grande Coulee falls as 800 feet in height (twice the height of Dry Falls) and almost 5 miles wide. Like Dry Falls, water here was substantially above the original lip of the falls.

Dry Falls, south of Coulee City, Washington, is the legacy of Ice Age Floods. Three miles long and 400 feet high, floodwaters were an estimated 300 feet above the present ground surface, meaning that instead of being a waterfall, this was only a ripple during the largest floods. Note grain elevator for scale.

Palouse Falls plummets 198 feet over four Columbia River basalt flows. Prior to Ice Age Floods, the Palouse River flowed in a canyon (Washtucna Coulee) to the north. The floods overtopped the drainage divide, carving a shortcut to the Snake River.

Where floods poured out of Grand Coulee onto the wide Columbia Plateau, they lost some velocity, and rock-bearing icebergs that had easily floated the deep waters of the Coulee were stranded in the slower, shallower water. South of Ephrata, they deposited a field of huge erratic boulders, known as the Ephrata Fan.

The floodwaters then scoured their way across the basaltic bedrock and soils of the Columbia Basin, creating multiple channels and pioneering new courses for streams. The Palouse River and Palouse Falls provide one excellent example. Palouse Falls is a scenic cataract where the Palouse River plunges 198 feet over three Columbia River basalt flows. Prior to the floods, the Palouse River flowed

down an entirely different canyon that is now the dry bed of gently meandering Washtucna Coulee. The energetic waters of one of the larger floods overtopped the divide between the river's original canyon and a smaller stream that fed the Snake River. They eroded several hundred feet of fragile loess that had formed a highland separating the two channels. The powerful floodwaters carved a new path for the Palouse River, which abandoned the old Washtucna channel in favor of a shorter, steeper, and far more spectacular route to the Snake.

Only a few dozen years separated each of the last glacial floods, based on the varves deposited in Glacial Lake Columbia. Although it took only a few days for floodwaters to drain Glacial Lake Missoula, several weeks were required for all the floodwaters to reach the Pacific.

For the largest Ice Age Floods, Wallula Gap acted as a hydraulic dam. The waters backed up, flooding the Pasco Basin, Quincy Basin, Yakima Valley, and even the Walla Walla area. The resulting lake, dubbed Lake Lewis (for Meriwether Lewis), formed to greater or lesser depths and durations with each large flood event. During the largest floods, water overtopped Wallula Gap. Pasco was underneath about 800 feet of water. In the Pasco Basin, only Candy, Badger, and Red Mountains would have kept their heads above water. Today, glacial erratics attest to the floodwater's height on these hills. Rhythmites and other sedimentary deposits define both the

Wallula Gap served as a "hydraulic dam," restricting the flow of floodwater and backing water up into a vast lake (Lake Lewis) that extended from Walla Walla to Yakima in the biggest floods. Water was almost 1,000 feet deep where Kennewick is today.

Near Walla Walla, Washington, a narrow canyon eroded into flood deposits allows geologists to date and count the number of large floods. Here, Richard Waitt of the U.S. Geological Survey first determined that at least forty large floods occurred between 17,000 and 15,000 years ago.

lake's extent and the floods that paused here. And perhaps most importantly, the sediment deposited in the Pasco, Quincy, and Walla Walla Basins by ponded floodwaters provide the rich soils that make the region a successful farming area.

Once the floodwaters entered the Columbia River, they eroded and widened the Columbia Valley's sidewalls and scoured the basaltic bedrock. In some places, including Crown Point, the waters overtopped the walls of the Columbia River Gorge, crafting side channels. In others their turbulence and high velocity literally bored holes in the rock that today appear as round ponds or depressions known as kolks.

Because rounding the sharp bend in the Columbia River at Portland slowed the currents, floodwaters backed up into the Willamette Valley. Portland was drowned beneath 300 feet of icy water in a temporary lake that backed up as far as Eugene. The turgid waters left a sort of "bathtub ring" of debris—including the mammoths—at a uniform elevation of about 300 feet all around the Willamette Valley. Erratic State Wayside west of Sheridan, Oregon, celebrates a huge boulder of argillite—a fine-grained, slightly metamorphosed rock—that was carried here from British Columbia aboard or embedded in an iceberg.

Another great flood that affected the Columbia Basin—and the Snake River Basin as well—was a single event known as the Bonneville Flood. Its volume was twice as great as the largest Missoula Flood. As glacial melting was in full swing in the Great Basin, 14,500 years ago, the enormous Lake Bonneville overtopped its northernmost confinement. At Red Rock Pass near Zenda, Idaho, the waters of the lake poured north, eroding the soft rocks of the pass, rushing downhill past the future site of Pocatello and into the Snake River.

Bonneville floodwaters broadened the canyon of the Snake across the Snake River

Washington's fertile Palouse consists of wind-blown silts carried from the north and west and deposited as undulating hills between Spokane and Walla Walla.

Plain, eroding and depositing these rounded chunks of canyon walls as "melon gravels" in oversized gravel bars. As the flood dropped into the steeper gradient of Hells Canyon, it accelerated. In the narrower parts of Hells Canyon, at Sinker Creek, flow rates topped 15 million gallons per second, or 0.3 cubic miles of water per second, one of the greatest flow rates known on the planet. (As a comparison, the total volume of material erupted from Mount St. Helens' 1980 event was 0.5 cubic miles.) This flood event continued for at least 8 weeks, and may have lasted as long as a year, before the turbulent waters eroded the pass's unconsolidated sediments down to harder bedrock at Red Rock Pass and water could no longer escape from Lake Bonneville into the Snake.

As the Ice Age Floods waned, winds generated near the front of retreating glacial ice again took charge of the landscape. A mantle of fine-grained silt coats most flood deposits, though they are far thinner than the fertile silts of the Palouse that have accumulated for as long as 2 million years. While Ice Age Floods transported much of the Palouse's soil into the Hermiston Basin, Yakima Basin, and Willamette Valley, Ice Age winds deposited the Palouse soils to begin with.

The Pleistocene Coast
Sea levels in the Northwest, and globally, declined almost 150 feet during the last glacial maximum, about 19,000 years ago, and as recently as 15,000 years ago were similarly low. As a consequence, winds toyed with marine sands, building them into huge dune fields. Vestiges of these dunes remain with us today along the southern and southcentral Oregon coast. Curt Peterson of Oregon State

University mapped out ten large dune sheets, including those of Oregon Dunes, and found that the dune tops, where dunes are stabilized inland, range from 140,000 to 46,500 years old. Peterson and colleagues found that prevailing onshore winds transported sands from the exposed continental shelf landward, to build huge dune fields. In fact, much of the sand in the older dunes was blown from now-submerged Heceta, Perpetua, and Stonewall Banks—which at the time were dry land that provided sand and even diverted coastal winds shoreward.

Humans arrived in the Northwest about 15,000 years ago during the Late Pleistocene, the time of extensive Ice Age Floods, well-developed lakes in the Basin and Range and High Desert, and sporadic volcanic activity. The question of whether any witnessed the floods, or were even swept away by them, whether they prowled the beaches and dunes of the last low stands of sea level, will likely remain unanswered. We do know that people inhabited the shorelines of the pluvial lakes by 15,000 years ago, thanks to work in Paisley Cave by Dennis Jenkins of the University of Oregon. The Northwest's landscape would be inhabited, groomed, and managed by humans from that time on. It was a new world, indeed.

The dunes along Oregon's coast were far more extensive during the Pleistocene when sea level was as much as 150 feet lower and winds more energetic. Today's dune fields, like these at Dellenback Dunes, are but a shadow of their former glory.

Northwest volcanoes threaten future eruptions. Here, a major rock avalanche produces an eruption-like plume in Mount St. Helens.

Global Temperatures

Millions of years | 2500 | PROTEROZOIC | 541 | PALEOZOIC | 252 | MESOZOIC | 66.0 | CENOZOIC

Camb | Ordo | Silu | Dev | Miss. | Penn | Perm | Trias | Juras | Creta | Paleo | Eoce | Olig | Mioc | Plioc | Pleist | Holo | Anth

CHAPTER 13 The Living Landscape *Pacific Northwest Geology in Action*

Today, the Northwest remains a place of geologic unrest, never satisfied with the status quo. It moves in time and tune with plate tectonics, responding like a dancer to the slight but persistent pressures of an ever-shifting planet. Ray Wells and Rick Blakely, both geophysicists with the U.S. Geological Survey, are watching this dance. Wells and Blakely track crustal movement by monitoring earthquakes, examining the strength and movement of past earthquakes, and using new technologies, including LIDAR (Light Imaging And Ranging) and sensitive GPS stations. They are especially interested in shallow earthquakes, which provide information about the direction and speed of the upper crust.

Their research shows that the western portion of the Northwest (Cascadia) and its small plates are caught between North America and the much larger Pacific plate, which is moving northwest at about 2 inches per year. As North America edges westward, the western United States, including Cascadia, is deforming over a broad area. This is what Wells and Blakely have deduced from gravity, seismicity, and magnetics:

1. The continental margin and the Juan de Fuca plate are breaking up into smaller crustal blocks that are being dragged northward by the motion of the Pacific plate.
2. Migrating continental margin terranes are creating three major Pacific Northwest crustal blocks: Washington, Oregon Coastal, and Sierra Nevada.
3. Vancouver Island and the Canadian Coast Mountains represent a relatively fixed buttress against which coastal terranes are deformed.
4. Coastal Oregon appears to be one large block, sometimes called a microplate; Washington appears to be several small, fault-bounded blocks.

In the Pacific Northwest's modern, unstable, and dynamic crust, faults, volcanoes, and earthquakes occur along block margins. The interior of the large Oregon coastal block is relatively quiet. However, in contrast, western Washington is seismically active, and earthquakes indicate north-south compression. The faults of the Seattle area relieve the stress of having Oregon's coast range shoved unceremoniously into Seattle's underbelly. The Yakima fold belt continues to expand. Volcanoes (such as the Cascades and the Boring Volcanic Field in the Portland area) are most abundant on the eastern edge of the rotating Oregon coastal block, a zone of crustal extension.

Tectonic Rotation

Perhaps the most thrilling ride that we are all getting on the plate-tectonic sleigh is the pronounced rotation of Oregon's coastal block as it shoves its way northward into Washington's underbelly, but also rotates northward and westward like a giant pendulum. More specifically, in the last 60 million years—since the time that its oldest rocks were formed as

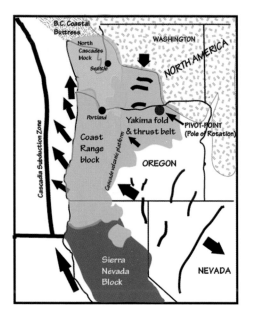

Tectonic Rotation. For at least the past 10 million years, the coastal ranges, Cascades, and Sierras have been pushed west by the opening Basin and Range. The result is tectonic rotation around a pole in northern Oregon.

oceanic lavas flows and seamounts—the Coast Range has rotated 60 degrees westward, or one degree about every million years. This may not seem like much to humans with our short lifespans, but it is heady stuff if you are a rock.

This rotation has compressed and uplifted the Oregon and Washington coast, the Willamette Valley, and the Puget Sound area, including Seattle. When the next great subduction zone earthquake releases the stress, the coast will rebound about two inches westward, and the upward bulge will flatten,

dropping the coast by as much as 3 feet, almost instantaneously. Few of us will enjoy this great and sudden tectonic shift, however. We will be too busy picking up the pieces of crumbled structures and inundated towns.

Within the past century, two volcanoes have produced significant eruptions—Mount Lassen and Mount St. Helens. Periodically others threaten—the bulge of mantle-derived helium and sulfuric flatulence on the west slopes of South Sister in 2003; the fumarolic breath of Mount Hood's Hot Rocks. Mount Hood erupted as recently as the 1790s, destroying forests on its western slopes, building Crater Rock dome, and dispatching destructive lahars down the Sandy River all the way to the Columbia. When William Broughton of the Vancouver Expedition explored the confluence of the Sandy River and Columbia in 1792, he noted the existence of a large sandbank that nearly blocked the Columbia. In 1805, Lewis and Clark made similar observations. This debris was the legacy of Hood's very recent eruptions and lahars.

Volcanic Hazards: Summit Collapse and Lahars

Hazards from the High Cascades are very real and should be of concern to all of us who live in their shadow. While Mount St. Helens' hazard is now obvious (almost to the point of being ignored, or mistakenly considered a "tamed" volcano that has already undergone a cataclysmic event and now poses little threat), many other High Cascade volcanoes

are considered active. These include Lassen, Shasta, South Sister, Bachelor, Hood, Adams, Rainier, Baker, and Glacier Peak. Hazards from the eruptions of these volcanoes include likely ash-fall and potentially devastating lateral blasts or pyroclastic flows. A pyroclastic flow erupted from Mount Hood into the White River Canyon about 1800 had a velocity of at least 85 mph according to Cameron and Pringle's calculations. And pyroclastic flows and lahars from eruptions between 1781 and 1793 traveled 50 miles down the Sandy River.

However, the most worrisome hazards from Cascade volcanoes are debris flows—either lahars or summit collapse. Lahars, spawned by sudden release of hot waters and rocks from near summit areas that follow river drainages, may inundate towns and dwellings under tens of feet of mud and rock. Major mudflows that have occurred within historic or near-historic Holocene time include the Sandy River debris, a lahar that followed the Sandy River all the way to the Columbia about 1794 and whose distal features were recorded by Lewis and Clark in 1805, and the Osceola Mudflow from Mount Rainier that buried the Puyallup Valley beneath hundreds of feet of mud and rock and pushed the shoreline of Puget Sound some 10 miles westward, creating 180 square miles of new land. Such events are not trivial. And they do not require an eruption to occur.

Summit collapse proffers an even greater catastrophe. Recognized as a hazard only after Mount St. Helens' north side vanished in a monstrous landslide that unleashed the eruption of 1980, epic summit collapse features have been documented at other Cascade peaks. The most egregious example is found northwest of Mount Shasta, where more than 50 square miles of the Shasta Valley were buried by debris when Shasta's ancestral cone collapsed about 300,000 years ago. Because sulfur-rich gases and associated sulfuric acid continually degrade the rock of volcanic summits, sudden and catastrophic collapse is an anticipated event on composite volcanoes. Many Cascade volcanoes, including Hood, Adams, and Rainier, have large areas of "rotten rock" in summit areas, making them prone to disaster.

And there is also a hazard from the sudden melting of glaciers. The highest concentration of lakes dammed by moraines in the conterminous United States is in the Mount Jefferson and Three Sisters Wilderness Areas in central Oregon, where there are currently eight moraine-dammed lakes. The largest, Carver Lake on South Sister, has a volume of almost 1 million cubic meters. Most of these lakes formed between 1920 and 1940 during a period of substantial warming and glacier retreat. In the Mount Jefferson and Three Sisters Wilderness Areas, there have been eleven different debris flows from four complete and seven partial drainings of moraine-dammed lakes. Most of these breaches occurred between 1930 and 1950, but some were as recent as the 1970s.

The Youngest Volcanoes

The Northwest's volcanic heart continues to beat. Within the past 10,000 years, the Cascades have sprouted new lava flows and

Basaltic vents similar to the eruptions that produced Diamond Craters may reappear in the Basin and Range and along the Brothers Fault Zone.

new eruptions well within the cultural memories of native peoples. Newberry produced the Big Obsidian Flow in AD 700, about the time that Constantine began his reign as Pope and Japan produced its first constitution. South Sister erupted about the same time, and has threatened again within the past decade.

In thin-crusted southeast Oregon, small geologically recent eruptions created noteworthy landscapes, including Diamond Craters and Jordan Craters about 6,000 years ago. Diamond Craters' earliest eruptions produced a series of craters or maars, including Twin Craters and Oval Craters, and small cinder cones (Big Bomb Crater, Red Crater). A second phase of activity produced a greater area of heated surface and additional maar (lava-poor, steam-driven) eruptions that produced deep pits and craters of the Central Crater complex, north of the present lava fields.

Diamond Craters' fluid olivine basalts exemplify many features of pahoehoe and shelly pahoehoe lava flows. At the most easily accessible crater, Lava Pit Crater, lava gutters spill ropy basalts into smaller pools with swirling textures. Hornitoes, also known as spatter cones, are vents where globs of fluid basalt are expelled from a blocked lava tube, building a structure up to about 30 feet high, displaying still-glassy lava spatters, drips, and drapes. Pressure ridges and inflation-collapse features abound on the barren and still fresh-looking basalt surfaces.

Cascadia Subduction Zone Earthquakes

Perhaps the greatest threat to Pacific Northwest communities lies in the certain recurrence of great (magnitude 9.0 and up) subduction zone earthquakes. Palpable evidence of these quakes—past and future—is found in sunken coastal marshes and estuaries up and down the Northwest coast. The wetland at Copalis, Washington, is noteworthy. A glance at the marshy meadowlands just north and east of downtown Copalis reveals mostly rushes, swampy ground, and a host of dead trees. But these trees are special. They are

Part of the compelling evidence for great subduction zone earthquakes every 300 years or so comes from the stumps of cedars drowned in 1700 when the last magnitude 9.0 subduction zone quake caused their lowland forest to subside beneath Pacific waves.

helped determine the Holocene history of great subduction zone quakes. Other trees killed as a result of quake-induced subsidence in coastal marshes include a ghost cedar forest on Long Island in Willapa Bay and drowned trees on several Columbia River islands upstream from Astoria and Point Deception.

A less poignant but more accessible patch of drowned forest lies along the beach at the south side of Sandy Point just off Bay Center Road on the northeast shore of Willapa Bay. They emerge from the sea's clammy grip only during low tides. Here, the drowned trees—spruce rather than cedar—are reduced to gnarly stumps that still grasp the sandy substrate—once a forest, now a beach. Onshore, just above the tide's reach, a few sandblasted relics still stand, tenuous reminders of the once-dark, moist forest doomed in a moment on January 26, 1700.

And we are due for the next great subduction zone earthquake. Evidence from sediment cores on the Astoria Fan, tsunami sand deposits in inland tidal zones, additional buried marshes, and buried sand liquefaction traces indicate that these quakes occur periodically, on average every 300 to 600 years, when the down-going seafloor unsticks itself and lurches powerfully forward. The next one could occur at any time.

Landslides: Geology on the Move
The Pacific Northwest's active geology entails more than just tectonics. The planet's epidermis (soils and surface) moves as

the remnants of a cedar forest, drowned more than 300 years ago by the last great subduction zone earthquake to shake the Northwest. The severe shaking compressed the ground in peat bogs and other locations with loose, organic-rich soils. The coastal and estuarine forests in these places sank, plunging the trees' roots below water level. Eventually, within a year or so of the earthquake, the trees that remained alive slowly died—including the trees at Copalis Marsh.

When USGS geologist Brian Atwater began trying to determine the date of this earthquake, he examined the huge cedar snags at Copalis Marsh. He discovered that they all died between the years 1690 and 1705. Further research and coordination with Japanese colleagues defined the date and even the time of the quake to 9 p.m., January 26, 1700. This date was determined from records of tsunamis generated by the quake that destroyed fishing vessels and coastal homes in Japan—where meticulous written records were kept by village shoguns. By determining the travel time back to the quake source off the Washington coast, geologists calculated the date and time of the earthquake with some precision. Since Atwater's defining work, other geologists have well—sometimes with tragic consequences. On the coast and the rainy west side of the Cascades, landslides and mass-wasting are forces to be reckoned with.

If you look north, across the Columbia River from Cascade Locks, Oregon, you'll see the reddish, raggedy countenance of Table Mountain. This is the headwall, or scarp, of the Northwest's largest recent landslide, the Bonneville Slide, also known as the Bridge of the Gods. The date of the Bonneville Slide is controversial. Dating by Nathan Reynolds, based upon both carbon-14 and the age of lichen that revegetated the barren landslide debris, suggests a date between 1670 and 1760. If this is correct, then the great Cascadia earthquake of January 26, 1700, may have triggered the slide. But newer research by Patrick Pringle and Jim O'Connor suggests a date of about 1450, which seems to correlate with nothing in particular—unless you heed the Klickitat Tribe's explanation for the slide.

In this story, Tyhee Saghalie—the Creator—traveled down the Columbia with his sons, Pahto and Wy'east. Upon reaching Hood River, Tyhee Saghalie gave his sons the choice of two places to call their own. Of course, the sons quarreled. To end the dispute, the Creator shot two arrows and instructed Pahto to follow the arrow north, to the site of Mount Adams. Wy'east settled where Mount Hood stands today. But when they both fell in love with Loowit (Mount St. Helens) they hurled rocks back and forth, and shook the earth violently, burying villages and forests, blocking the river with a great slide. This explanation suggests that an eruption of one or more volcanoes coincided with the slide, although no eruption here is known (yet) from either time indicated by radiometric dates.

What *is* certain is that the massive slide blocked the Columbia River, creating a dam about 200 feet high and more than 3 miles long. The resulting lake backed up almost 40 miles nearly to Hood River, Oregon. It may have been a week or more before the Columbia overtopped the dam, and eventually removed most of it. The slide's footprint created rough waters that Lewis and Clark dubbed the "Cascades" of the Columbia. The range of volcanoes we know as the Cascades was named after this uneasy water.

The Bonneville Slide still moves today, and is one of many landslides that disturb roads and disrupt lives. In 1999, landslides caused the destruction and condemnation of 138 homes in Kelso, Washington. A landslide first noticed in 1996 in The Capes development on Oregon's central coast has since led to the condemnation of more than 20 houses. In March 2014, a mudslide blocked the Stillaguamish River, engulfed 49 homes, and killed 42 people in Oso, Washington. In the Pacific Northwest, landslides cause, on average, about two million dollars' worth of damage each year, according to state geological surveys. In the past 50 years, more than 200 deaths have occurred due to landslides, including the horrific 2014 mudslide in Oso, Washington. Like many slides, the Oso slide had been mapped. Geologists had warned of slope instability. But this, tragically, had little impact on people who wished to live along a scenic river, or on local governments who should have heeded geologists' warning and mapping and established building zones and codes to avert needless human disaster.

The Anthropomorphic Northwest

James Hutton, regarded as the father of modern geology, observed that the present is the key to the past—that geological processes we see today also formatted past landscapes. Another concept, just as important, is that the past is the key to the future.

Today, we are building new crust through the action of Cascade volcanoes, and rending it apart through tectonic rotation. Landscapes uplifted by faulting and plate convergence tend to slide downslope, restoring geologic balance while creating human chaos. We await the next Cascadia quake, or the next volcanic tantrum, while the steady beat of more patient processes continues. The Columbia builds its offshore fan, albeit with far less sediment than before we built dams. Winds sculpt dunes on the Oregon coast, though at a different pace than before off-highway vehicles. Rocks decay into clay and then to soil, but more readily than before acid rain. We have arrived at the Anthropocene, a time period in which humans are a geological force. This is a factor that Hutton, who traipsed the Scottish moors in the mid-eighteenth century, probably never anticipated. But as we move toward the future, perhaps the most important geological consideration is us.

As climates change, the Northwest will see rising snow levels, faster runoff, drier summers, and overall lower stream flows, as well as generally warmer weather. Deserts, such as the Alvord, will become drier and less hospitable.

CHAPTER 14   Tomorrow   *Changing Climates, Moving Plates*

An important lesson from the Northwest's active geologic record is that neither landscapes nor climates are stable. Throughout the entire history of the Northwest, landscapes and ecosystems have changed. Oceanic and continental configurations have shifted. Solar influx and Earth's orbit change, usually in cyclical patterns. And because of these factors, the globe has varied repeatedly from a snowball to a greenhouse during at least the past 700 million years. Extinctions have been handmaidens to climate change.

In the Northwest, we have scant records of Snowball Earth, 700 million years ago, and the planet's rescue by volcanism and CO_2. Our exotic terranes pretty much skip over the angst of climate-driven extinctions in the Ordovician and Devonian, the oxygen-rich coal swamps and ensuing carboniferous glaciations as atmospheric CO_2 plummeted. And although Permian rocks are ensconced in the Blue Mountains and the Klamaths, for us, the page that contains the Permian extinction,

replete with volcanic fury, toxic air, and an acid ocean that extinguished 95 percent of planetary life, has mercifully been omitted. Similarly, a record of the CO_2-driven extinction at the end of the Triassic, 201 million years ago, is absent here. Our Cretaceous rocks include several chapters but omit dinosaurs almost entirely, and leave the entire meteorite impact tragedy to the record elsewhere.

But what we *do* have here is an astoundingly detailed and unequivocal record of climate change in the Tertiary, once the dinosaurs were gone. We are the Gutenberg Bible of the Paleocene and Eocene and their tropic climates. The Painted Hills record a dwindling CO_2 concentration with time through the Oligocene, and a cooling atmosphere. Right here in our backyard, we have evidence for the mid-Miocene thermal optimum, which coincides precisely with the erupting of our own Columbia River basalts. And of course, there are the cooling but willy-nilly climates of the Pliocene and the multiple advances and retreats, cold and warmth, of the Pleistocene.

So there should be no doubt among those familiar with geologic history that climates change, and that atmospheric composition, as well as other factors of oceanic configuration, solar influx, and volcanic perturbations, are major contributors. When atmospheric carbon dioxide and/or methane increase, global climates, especially those at the poles, get warmer. When atmospheric greenhouse gas decreases, the globe may plunge into an ice age. It is this record that climate change "skeptics" seem to forget, or are ignorant of in the first place.

Perhaps the best and most irrefutable example of rapid atmospheric-driven climate change occurred at the Paleocene-Eocene boundary, recently dated with precision as 55.728 million years ago. This date does not coincide with any calculated or expected close proximity to the sun. Rather, it coincides with a time in the 400,000-year orbital eccentricities when the Earth would have been far—and getting farther—from the Sun. But it does

In August 2012, smoke from Siberian forest fires drifted toward the Northwest, producing record ozone and smog levels for Bellingham and Seattle and obscuring much of the North Cascades, including Mount Shuksan.

coincide with a scathing burst of hotspot (Faroe Island/Mid-Atlantic Ridge) volcanic activity, and, based on oceanic cores, a spike in both atmospheric methane and atmospheric carbon dioxide. This increase is the fastest known in the geologic record. Ultimately, during the perhaps 200,000-year length of this warm episode, atmospheric CO_2 reached as high as 1,200 ppm. Temperatures rose, precipitation changed. Oceans acidified. And even though this climate "excursion" did not mark a major extinction, it hit the reset button for mammalian reproduction. Marsupials declined precipitously; placentals (us) gained ground.

This and other changes in atmospheric composition through geologic time hold lessons for us today. We have irrefutable data that atmospheric CO_2 is rising. Nowhere on the horizon are there volcanic eruptions or other natural sources sufficient to make a significant contribution. The source of greenhouse gas is us. What is more alarming is the rate of increase. In the Paleocene, atmospheric carbon dioxide emissions peaked at between 300 million and 1.7 billion metric tons per year. That seems like a lot. But it pales in comparison with the rate that humans are producing CO_2—estimated, in 2010, at 8.8 billion metric tons per year, half that enters the atmosphere and remains. This rate is due to increase as China, India, and other nations "modernize." Atmospheric CO_2, measured at Mauna Loa Observatory, has climbed to more than 400 ppm in May 2013. It was 370 ppm in 2000.

At this rate, in a little more than a century, we will achieve what hard-working volcanoes required about 20,000 years to do: raise Earth's atmospheric CO_2 to 600 ppm—a level that had bananas growing in central Oregon and palm trees prospering in Bellingham.

We can also examine a time closer to us—the Pliocene—when carbon dioxide levels flirted with 600 ppm before fading to 300 ppm or less. Oceans and continents were aligned much as they are today. We cannot blame climate on errant ocean currents or weird clusters of landmasses or solar proximity. Today we seem to experience increasingly severe weather events. Hurricanes have rearranged cities and cost trillions in repair and reconstruction—which, one might argue, are bad investments in the face of rising sea levels. Typhoons and floods have devastated Thailand, the Philippines, and other parts of Asia. (As Bill McKibben wryly noted at a talk at the Geological Society of America after Superstorm Sandy, "If God wants us to do something about climate change, His aim is getting better.")

In the Pliocene's record, we also see clear evidence of monumental storms during unsettled times of warming. In the Quileute Formation, on Washington's northwest coast, storm-generated deposits called tempestites document increased intensities of storms, waves, and river-flood plumes during times of warming and sea level rise. The Pliocene warm periods were times when the Northwest coast bore the brunt of severe storms.

Whether that will be the consequence of today's climate change is uncertain. But it certainly was a consequence of the past.

If the prospect of an ice-free Arctic for the first time in 34 million years leaves you undaunted, and perhaps even enthused about the commercial prospects of the Northwest Passage or access to new resorts on the pristine Arctic Ocean Shore, and even more oil, then consider rapid climate change's demonstrated effect on life. The Paleocene-Eocene boundary, when global CO_2 was, by some estimates, 1,200 ppm, produced major oceanic acidification, driving about 35 percent of planktonic species to extinction. On land, we lost several orders of animals, including most marsupials. Throughout the geologic record, rapid climate shifts have triggered extinction events.

The question for us is how far down the path to the Paleocene do we wish to travel? (Or for that matter, how far down the path to the end-Permian?) How much global flora and fauna, marine and terrestrial, can we afford to lose? Should we lose? And what will happen as ecosystems unravel while they attempt to adjust? Humanity may not be threatened, but much of the wildlife and plants that live now are. These are real-world questions, not academic ones. And the likelihood of unintended consequences is troubling.

Although it seems easier to endure, to turn up the air conditioning, or build a higher sea wall, rather than change global economies,

Another magnitude 9.0 or greater earthquake is certain in the Northwest. It will likely unleash a tsunami that could devastate the Pacific coast. Pictured, Japanese dock—debris from the March 11, 2011, magnitude 9.0 Honshu, Japan, earthquake and tsunami, which washed up on Oregon shores a year later.

the geologic record is clear that irreversible changes—and not for the better—lie ahead. Humans, with life spans approaching a century, still consider a hundred years a long time, and a millennium too distant for planning. Yet our actions today require us to think on a geologic time scale, for humans have become agents of geologic-scale global change. As we venture into an uncertain future, it is well to keep the lessons of the past in mind.

If we take no other lesson from geology, it should be this: change is constant. Change at every pace and every scale. Continents shift and mountains rise on multiple scales. Earthquakes dispense a billion joules of energy in a single-second fracture of stone. Tsunamis that travel for minutes end with effects that are felt for decades. Motion at a fingernail's growth rate raises the Himalayas 5 miles in 30 million years. This is never the same planet. Each day is new. Each day, North America has moved almost imperceptibly

farther northwest. Each day, the ocean gyres modify their precise velocity and path. Each high tide, each low tide, is unique. Each day the planetary axis wobbles. Each day, the atmosphere changes. No day or moment is the same. We should cherish them all, past, present, and future. Most important of all, this is the Anthropocene, and the future is in our hands.

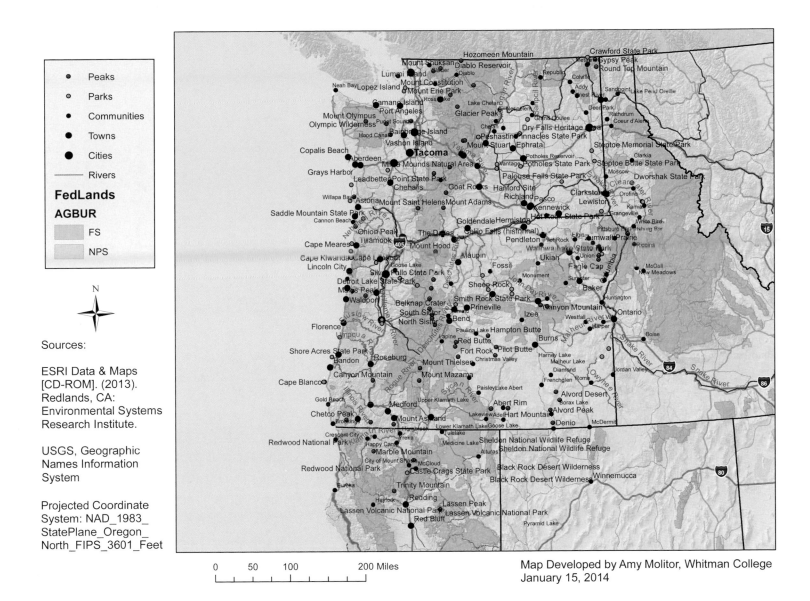

Sources:

ESRI Data & Maps
[CD-ROM]. (2013).
Redlands, CA:
Environmental Systems
Research Institute.

USGS, Geographic
Names Information
System

Projected Coordinate
System: NAD_1983_
StatePlane_Oregon_
North_FIPS_3601_Feet

Map Developed by Amy Molitor, Whitman College
January 15, 2014

Throughout this book, ages, temperatures, atmospheric concentrations of oxygen and carbon dioxide, as well as the locations of ancient continents and other quantifications appear frequently. Determining these parameters for ecosystems millions and even billions of years old seems like magic, or perhaps as though geologists are just making up convenient numbers.

Nothing could be further from the truth. New technologies and instrumentation have allowed us to probe the conditions of the past with considerable accuracy. This section of the book describes only a few of these technologies. Here is a very brief summary of how we determine age, temperatures/climate, atmospheric composition, and location.

Determining Age

The age of geologic materials (notably of rocks) is determined using radiometric dating—the relative abundance of parent radioactive elements and their decay products, or "daughter" elements. At the time a mineral is formed, the parent element is incorporated, as part of a mineral lattice. With time, that element decays into its daughter elements. Using decay constants (the rate at which the parent decays into the daughters), it is possible to determine how long that element has been contained in the rock—and hence, how old the rock is. To determine the age of a rock, either the whole rock, or, increasingly, individual minerals, are analyzed for radioactive elements and the concentration of related daughter decay products. From this,

geologists can calculate the length of time since the rock or mineral formed.

Many isotopic "systems" are used for these calculations. Exactly which is best to utilize depends upon which elements are present in the rock, the half-lives of the radiogenic isotopes to be measured, and also the estimated or anticipated age of the rock. For example, the half-life (time required for half of the parent element to decay into the daughter element) of Uranium-238 (U^{238}) in its decay to Lead-206 (Pb^{206}) is about 4.5 billion years. Rocks that are between about 3 billion years and 2 million years contain sufficient amounts of both the parent U^{238} and daughter Pb^{206} to accurately measure and determine an age. But in very young rocks—for example, an andesite erupted from Mount Rainier's main cone—the amount of Pb^{206} produced by decay is not yet sufficient to measure accurately. Especially for young volcanic rocks, determining an accurate and precise age can be challenging.

The systems most commonly used for radiometric dating or geologic materials are uranium-lead, samarium-neodymium, potassium-argon, rubidium-strontium, uranium-thorium, argon-argon, lutetium-hafnium, and rhenium-osmium. Importantly, note that carbon-14 dating is not used for geological materials because few rocks contain carbon and the half-life of carbon is short, making dating anything older than about 50,000 years very iffy. The exception to this is dating young lava flows, especially in the High Cascades. And it's not the actual rock that is dated. Instead, geologists look

for the casts of trees burned by the lava flow and collect charcoal from the tree casts/wells to date the time that the tree burst into flame.

Why do geologic dates for important events (the Permian extinction, for example) seem to change? First, as our instrumentation becomes more sophisticated, we are able to measure parent-daughter abundances more precisely, and thus determine more accurate dates. Second, geochronologists keep fine-tuning a very important number: the decay constant. This tells us the rate at which a parent transforms into a daughter (isotope). Change the decay constant and you have changed the length of time it takes to produce a daughter isotope. As decay constants are refined, dates change—sometimes radically. A decade ago, for instance, the decay constant for C^{14} was redetermined, and overnight, many dates changed—such as the eruption of Mount Mazama (Crater Lake), which was suddenly 7,700 years ago rather than 6,500 years ago. This has nothing to do with errors in previous dates. It has everything to do with greater precision. Science marches on.

Determining Temperatures/Climate

How can we say with authority what the temperature of the Arctic Ocean was during the Paleocene, 55 million years ago? Or what the average daily temperature was on an average day in the Oligocene? The answer lies in oxygen—specifically, stable oxygen isotopes.

Like all elements, oxygen is made up of a nucleus of protons and neutrons, surrounded by a cloud of electrons. Oxygen comes in heavy and light varieties, or isotopes, which are useful for paleoclimate research. "Light" oxygen-16 is the most common isotope found in nature, followed by much lesser amounts of "heavy" oxygen-18. Oxygen-16 will evaporate more readily than oxygen-18 since it is lighter. Hence, during a warm period, the relative amount of oxygen-18 will increase in the ocean waters since more of the oxygen-16 is evaporating. This same ration is maintained in biogenic ocean and freshwater sedimentary rocks including limestone and chert, and in minerals that form using seawater. Hence, water temperatures of the past can be determined quite accurately by measuring oxygen isotopes in sedimentary rocks and minerals, as well as glacial ice.

Determining Atmospheric CO2 and the Carbon Cycle

How can geologists determine the amount of carbon dioxide in the atmosphere 300 million years ago? Or in 1700? The answer lies in stable isotopes of carbon.

Three isotopic methods, or "proxies," are commonly applied to geologic materials to gauge ancient carbon dioxide. All are based on the partitioning of the two stable isotopes of carbon: C^{12} and C^{13}. Plants take up C^{13} more slowly than C^{12}. Consequently, organic matter is depleted in C^{13} proportional to the amount of atmospheric carbon dioxide. But, when atmospheric CO2 levels are low, plants don't distinguish as much between the two isotopes compared with times when atmospheric carbon dioxide is high. The ratio of C^{13} to C^{12} in sedimentary rocks is proportional to atmospheric CO2 at the time the rocks were deposited.

First, when atmospheric carbon dioxide is low, the $C^{13/12}$ ratio in calcium carbonate and other minerals that have formed from atmospheric carbon

is elevated. This method requires using calcium carbonate (calcite) or the hydrous iron carbonate goethite formed in ancient soils. Geologists can also determine ancient levels of CO2 using the $C^{13/12}$ ratio in phytoplanktonic organic remains—notably, hydrocarbons or the boron isotopes found in planktonic foraminifera. Why boron? Boron uptake by plankton is regulated by the pH of seawater. And seawater pH is determined by the atmospheric concentration of carbon dioxide, which forms carbonic acid.

One additional proxy is widely used where adequate preservation of fossil leaves permit: the stomatal index. Plants inhale carbon dioxide through the stomata in their leaves. When atmospheric CO2 is abundant, plants need to inhale less strenuously to supply their metabolic needs, and thus develop few stomata on their leaves. When CO2 is scarce, leaves bear more/larger stomata. Studies of well-preserved fossil leaves revealed a linear relationship—the higher the CO2, the fewer stomata. This is actually a quantifiable relationship that persists today, according to Dana Royer at Penn State, Robert Berner of Yale, and colleagues.

Determining Paleolatitude

How can we determine where continents and plates were located millions of years ago? When igneous rocks cool and solidify, iron oxide minerals such as magnetite (the original compass needle) align themselves with the Earth's magnetic field. They accomplish this in three dimensions so their orientation reveals not only which way was north, but also how far north or south of the equator they were. Geologists (paleomagnetists) collect carefully oriented samples, usually as a core extracted from a stable outcrop, and use sensitive magnetometers to measure the precise orientation of the magnetite within their samples.

A U.S. Geological Survey geologist samples a basalt outcrop to take the magnetic bearing of the basalt as it formed.

Well-preserved fossil leaves that retain fine structures and stomata provide a record of past atmospheric carbon dioxide.

REFERENCES AND READING

Noteworthy Books About PNW Geology

Alt, D. D. 2001. *Glacial Lake Missoula and its Humongous Floods.* Missoula, MT: Mountain Press Publishing Company.

Bishop, E. M. 2004. *In Search of Ancient Oregon.* Portland, OR: Timber Press.

Harris, S. 2005. *Fire Mountains of the West.* Missoula, MT: Mountain Press.

Lilly, Robert J., 2005. *Parks and Plates: The Geology of Our National Parks, Monuments, and Seashores.* New York, NY: W.W. Norton.

Miller, M. Forthcoming. *Roadside Geology of Oregon.* Missoula, MT: Mountain Press.

Orr, E., and W. Orr. 2012. *Oregon Geology.* Corvallis, OR: Oregon State University Press.

———. 2006. *Geology of the Pacific Northwest.* Long Grove, IL: Waveland Press.

Tabor, R. W., and R. A. Haugerud. 1999. *Geology of the North Cascades: A Mountain Mosaic.* Seattle, WA: The Mountaineers Books.

Vallier, T. L. 1998. *Islands and Rapids: A Geologic Story of Hells Canyon.* Lewiston, ID: Confluence Press.

Websites

GENERAL PNW GEOLOGY

Burke Museum: www.burkemuseum.org/geology
California Geological Survey: www.conservation.ca.gov/cgs/Pages/Index.aspx
Digital Geology of Idaho: geology.isu.edu/Digital_Geology_Idaho/
Idaho Geological Survey: www.idahogeology.org/
Idaho Museum of Natural History: imnh.isu.edu/home/
Klamath Tribal History, Giiwaas (Crater Lake): klamathtribes.org/background/giiwaas.html

Oregon Department of Geology and Mineral Industries: www.oregongeology.org/sub/default.htm
Washington Geological Survey: www.dnr.wa.gov/ResearchScience/Topics/GeologyofWashington/Pages/geolofwa.aspx

VOLCANOES

USGS, Cascades Volcano Observatory: volcanoes.usgs.gov/observatories/cvo/
USGS, Columbia Plateau: www.nature.nps.gov/geology/usgsnps/province/columplat.html
USGS, Volcanoes of U.S.: www.nature.nps.gov/geology/usgsnps/province/cascade2.html

EARTHQUAKES/SEISMICITY

Pacific Northwest Seismic Network: www.pnsn.org/
USGS Seismic Hazards, PNW: earthquake.usgs.gov/regional/pacnw/

PLATE TECTONICS

Christopher Scotese PaleoMap Project: www.scotese.com
Ron Blakey Colorado Plateau Systems Paleomaps: cpgeosystems.com/paleomaps.html

ICE AGE

Ice Age Floods Institute: www.iafi.org/floods.html
www.hugefloods.com

Geologic Field Trip and Hiking Guides

Babcock, R. S., and R. J. Carson. 2000. *Hiking Washington's Geology.* Seattle, WA: The Mountaineers Books.

Bishop, E. M. 2006. *Hiking Oregon's Geology.* 2nd edition. Seattle, WA: The Mountaineers Books.

Bjornstad, B. 2006. *On the Trail of the Ice Age Floods: A Geological Field Guide to the Mid-Columbia Basin.* Sandpoint, ID: Keokee Books.

Bjornstad, B., and E. Kiver. 2012: *On the Trail of the Ice Age Floods: The Northern Reaches.* Sandpoint, ID: Keokee Books.

Carson, R. J., and K. R. Pogue. 1996. *Flood Basalts and*

Glacier Floods: Roadside Geology of Parts of Walla Walla, Franklin, and Columbia Counties, Washington: Olympia, WA: Washington Division of Geology and Earth Resources Information Circular 90.

Halliday, W. R. 1963. *Caves of Washington*: Olympia, WA: Washington Division of Mines and Geology Information Circular 40.

Joseph, N. L., et al., eds. 1989. *Geologic Guidebook for Washington and Adjacent Areas.* Washington Division of Geology and Earth Resources Information Circular 86.

O'Connor, J. E., R. J. Dorsey, and I. Madin. 2009. *Volcanoes to Vineyards: Geologic Field Trips through the Dynamic Landscape of the Pacific Northwest.* Field Guide 15. Denver, CO: Geological Society of America.

Swanson, T. 2005. *Western Cordillera and Adjacent Areas.* Field Guide 4: The Geological Society of America.

Important/Interesting Classic and Recent Scientific Papers

Barry, T. L., S. Self, S. P. Kelley, P. Hooper, and M. Widdowson. 2010. "New ^{40}Ar/^{39}Ar dating of the Grande Ronde lavas, Columbia River Basalts, USA; Implications for duration of flood basalt eruption episodes." *Lithos* 118: 213–222.

Baker, V. R. 2009. "The Channeled Scablands: A Retrospective." *Annual Reviews, Earth and Planetary Science* 37: 393–411.

Burgess, S. D., S. Bowring, and S. Shen. 2014. "High-precision timeline for Earth's most severe extinction." *Proceedings of the National Academy of Sciences* 111: 3316–3321.

Camp, V. E., M. E. Ross, R. A. Duncan, N. Jarboe, R. S. Coe, B. B. Hanan, and J. A. Johnson. 2013. "The Steens Basalt: Earliest lavas of the Columbia River Group." *Geological Society of America Special Papers* 497: 87–116.

Cummings, M., J. Evans, M. Ferns, and K. Lee. 2000. "Stratigraphic and structural evolution of the middle Miocene synvolcanic Oregon-Idaho graben." *Geologic Society of America Bulletin* 112: 668–682.

Dickinson, W. R. 2005. "Evolution of the North American Cordillera." *Annual Review of Earth and Planetary Sciences* 32: 13–45.

Dorsey, R. J., and T. A. LaMaskin. 2008. "Mesozoic collision and accretion of oceanic terranes in the Blue Mountains province of northeastern Oregon: New insights from the stratigraphic record." *Arizona Geological Society Digest* 22: 325–332.

Evans, D., and R. Mitchell. 2011. "Assembly and breakup of the core of Paleoproterozoic-Mesoproterozoic supercontinent Nuna." *Geology* 39: 443–446.

Evarts, R., R. Conrey, R. Fleck, and J. Hagstrum. 2009. "The Boring Volcanic Field of the Portland-Vancouver Area, Oregon and Washington: Tectonically anomalous forearc volcanism in an urban setting." *GSA Field Guide* 15: 253–270.

Fierstein, J., W. Hildreth, and A. Calvert. 2011. "Eruptive history of South Sister, Oregon Cascades." *Volcanology and Geothermal Research* 207: 145–179.

Finnegan, S., N. A. Heim, S. E. Peters, and W. W. Fischer. 2012. "Climate change and the selective signature of the Late Ordovician mass extinction." *Proceedings of the National Academy of Sciences* 109: 6829–6834.

Gains, R. R., E. U. Hammarlund, X. Hou, C. Qi, S. E. Gabbott, Y. Zhao, J. Peng, and D. E. Canfield. 2012. "Mechanism for Burgess Shale-type preservation." *Proceedings of the National Academy of Sciences* 109: 5180–5184.

Gaschnig, R., J. D. Vervoort, R. S. Lewis, and B. Tikoff. 2011. "Isotopic evolution of the Idaho Batholith and Challis Intrusive Province, Northern U.S. Cordillera." *Journal of Petrology* 52: 2397–2429.

Hildreth, W., and M. Lanphere. 1994. "Potassium-Argon geochronology of a basalt-andesite-dacite arc system: The Mount Adams Volcanic Field, Cascade Range of southern Washington." *Geologic Society of America Bulletin* 106: 1413–1429.

Hoffman, P. F., A. J. Kaufman, G. P. Halverson, D. P. Schrag. 1998. "A Neoproterozoic Snowball Earth." *Science* 281: 1342–1346.

Hooper, P., and J. Hawkesworth. 1993. "Isotopic and geochemical constraints on the origin and evolution of the Columbia River Basalt." *Journal of Petrology* 34: 1203–1246.

Hooper, P. R., V. E. Camp, S. Reidel, and M. E. Ross. 2002. "The origin of the Columbia River flood basalt province: Plume versus nonplume models." *Geologic Society of America Special Papers* 430: 635–668.

Jaraula, C., K. Grice, R. J. Twitchett, M. E. Böttcher, P. LeMetayer, A. G. Dastidar, and L. F. Opazo. 2013. "Elevated pCO2 leading to Late Triassic extinction, persistent photoic zone euxinia, and rising sea levels." *Geology* 41: 955–558.

Keleman, P. B., J. Matter, E. E. Streit, J. F. Rudge, W. B. Curry, and J. Blusztajn. 2011. "Rates and mechanisms of mineral carbonation in peridotite: Natural processes and recipes for enhanced, in-situ CO2 capture and storage." *Annual Reviews of Earth and Planetary Sciences* 39: 545–576.

Kuckenberg, S., D. Whitney, M. Fanning, R. Teyssier, and J. Dunlap. 2008. "Paleocene-Eocene migmatite crystallization, extension, and exhumation of the northern Cordillera: Okanogan dome, Washington, USA." *Geologic Society of America Bulletin* 120: 912–929.

Kurschner, W., Z. Kvack, and D. Dilcher. 2008. "The impact of Miocene atmospheric carbon dioxide fluctuations on climate and the evolution of terrestrial ecosystems." *Proceedings of the National Academy of Sciences* 102: 449–453.

LaMaskin, T. A., J. D. Vervoot, R. J. Dorsey, and J. E. Wright. 2011. "Early Mesozoic paleogeography and tectonic evolution of the Western United States: Insights from detrital zircon U-Pb geochronology, Blue Mountains Province, northeastern Oregon." *Geologic Society of America Bulletin* 123: 1939–1965.

Lindsley-Griffin, N., J. Griffin, and J. Farmer. 2008. "Paleogeographic significance of Ediacaran cyclomedusoids in the Antelope Mountain Quartzite, Yreka Subterrane, eastern Klamath Mountains, California." *Geologic Society of America Special Papers* 442: 1–34.

Lui, L., and D. Stegman. 2012. "Origin of Columbia River Flood Basalt controlled by propagating rupture of the Farallon slab." *Nature* 482: 386–389.

McInerney, F. A., and S. L. Wing. 2011. "The Paleocene-Eocene Thermal Maximum: A perturbation of carbon cycle, climate, and biosphere, with implications for the future." *Annual Reviews of Earth and Planetary Science* 39: 489–516.

Metzger, E., R. Miller, G. Harper. 2002. "Geochemistry and tectonic setting of the Ophiolitic Ingalls Complex, North Cascades, Washington: Implications for correlations of Jurassic Cordilleran Ophiolites." *The Journal of Geology* 110: 543–560.

Morris, G. A., P. B. Larson, and P. R. Hooper. 2000. "Subduction style" magmatism in a non-subduction setting: The Colville Igneous Complex, NE Washington State, USA." *Journal of Petrology* 41: 43–67.

O'Keefe, F. R., and L. M. Chiappe. 2012. "Viviparity and K-selected life history in a Mesozoic marine plesiosaur (Reptilia, Sauropterygia)." *Science* 333: 870–873.

Pavlis, G., K. Sigloch, S. Burdick, M. Fouch, and M. Vernon. 2012. "Unraveling the geometry of the Farallon Plate: Synthesis of three-dimensional imaging results from USArray." *Tectonophysics* 535: 82–102.

Payne, J., and M. Clapman. 2012. "End-Permian mass extinction in the oceans: An ancient analog for the twenty-first century?" *Annual Review of Earth and Planetary Sciences* 40: 89–111.

Reidel, S., V. Camp, T. Tolan, and B. Martin. 2013. "The Columbia River flood basalt province: Stratigraphy, areal extent, volume, and physical volcanology." *Geologic Society of America Special Papers* 497: 1–42.

Reidel, S., and T. Tolan. 2013. The Late Cenozoic evolution of the Columbia River system. *Geologic Society of America Special Papers* 497: 201–230.

Reiners, P. W., T. A. Ehlers, J. I. Garver, S. G. Mitchell, D. R. Montgomery, J. A. Vance, and S. Nicolescu. 2005. "Late Miocene exhumation and uplift of the Washington Cascade Range." *Geology* 30: 767–770.

Renne, P. R., A. L. Deino, F. J. Hilgen, K. F. Kuiper, D. F. Mark, W. S. Mitchell III, L. E. Morgan, R. Mundil, J. Smit. 2013. "Time scales of critical events around the Cretaceous-Paleogene Boundary." *Science* 339: 684–687.

Retallack, G. 2007. "Cenozoic climate change on land." *Journal of Geology* 115: 271–294.

Royer, D., R. Berner, I. Montenaz, N. Tabor, and D. Beerling. 2008. "CO2 as a primary driver of Phanerozoic climate." *GSA Today* 14: 4–10.

Salzmann, U., A. M Haywood, and D. J. Lunt. 2009. "The past is a guide to the future? Comparing Middle Pliocene vegetation with predicted biome distributions for the twenty-first century." *Philosophical Transactions of the Royal Society A* 367 (1886): 189–204.

Schmidt, M., and A. Grunder. 2009. "The evolution of North Sister: A volcano shaped by extension and ice in the central Oregon Cascade Arc." *Geologic Society of America Bulletin* 121: 643–662.

Schwartz, J., A. Snoke, C. Frost, C. Barnes, L. Gromet, and K. Johnson. 2009. "Analysis of the Wallowa-Baker terrane boundary: Implications for tectonic accretion in the Blue Mountains province, northeastern Oregon." *Geologic Society of America Bulletin* 121: 221–236.

Shaller, M. F., J. D. Wright, and D. V. Kent. 2011. "Atmospheric PCO2 perturbations associated with the Central Atlantic Magmatic Province." *Science* 331: 1404–1409.

Sheehan, P. M. 2001. "The Late Ordovician mass extinction." *Annual Review of Earth and Planetary Sciences* 29: 331–364.

Sisson, T. W., V. J. M. Salaters, and P. B Larson. 2014. "Petrogenesis of Mount Rainier andesite: Magma flux and geologic controls on the contrasting differentiation styles at stratovolcanoes of the southern Washington Cascades." *Geological Society of America Bulletin* 126: 122–144.

Smith, G., N. Morgan, and E. Gustafson. 2000. "Fishes of the Mio-Pliocene Ringold Formation, Washington: Pliocene capture of the Snake River by the Columbia." *The University of Michigan Museum of Paleontology* Bulletin 32.

Stromberg, C. E. A. 2011. "Evolution of grasses and grassland ecosystems." *Annual Review of Earth and Planetary Sciences* 39: 517–544.

Sun, Y., M. Joachimski, P. Wignall, C. Yan, Y. Chen, H. Jiang, L. Wang, and X. Lai. 2012. "Lethally hot temperatures during the Early Triassic Greenhouse." *Science* 338: 366–370.

Sutcliffe, O. E., J. A. Dowdeswell, R. J. Whittington, J. N. Theron, and J. Craig. 2013. "Calibrating the Late Ordovician glaciation and mass extinction by the eccentricity cycles of Earth's orbit." *Geology* 28: 967–970.

Walker, J. D., J. W. Geissman, S. A. Bowring, and L. E. Babcock. 2013. "Geological Society of America Time Scale." *Geological Society of America Bulletin* 125: 259–272.

Young, G. A. 2010. "Placoderms (Armored Fish): Dominant vertebrates of the Devonian period." *Annual Reviews of Earth and Planetary Sciences* 38: 523–550

Zachos, J. C., G. R. Dickens, and R. E. Zeebe. 2008. "An early Cenozoic perspective on greenhouse warming and carbon cycle dynamics." *Nature* 451: 279–283.

aa: A common type of basaltic lava flow in which the top and bottom of the flow are sharp-edged rubble, although the middle remains solid.

amphibole: A family of usually iron-rich silicate minerals common in igneous and metamorphic rocks. Hornblende is one type of amphibole.

amphibolite: A high-grade metamorphic rock composed largely of the dark, iron-rich mineral hornblende.

andesite: A gray volcanic rock typical of composite volcanoes, including most of the Cascades. In outcrop, the rocks display platy jointing, and may resemble shale.

anticline: A fold with limbs that continue downward, away from the fold axis. It may be easier to think of it as an "up-fold."

argillite: A fine-grained, silica-rich sedimentary rock composed of fine clays.

arkose: A sandstone that usually contains at least 25 percent feldspar, has a mineralogical composition similar to granite, and is usually weathered from granitic rocks close to the depositional site.

ash: The fine rock particles erupted from volcanoes, composed of very rapidly cooled volcanic rock, which thus are often tiny shards of glass.

ash-flow tuff: A rock consisting of volcanic ash deposited by a hot blast from a volcano. See also *ignimbrite*.

basalt: A fine-grained, dark brown to black rock on a freshly broken surface. It derives its dark color from its high content of iron and magnesium and subsequent abundance of minerals rich in iron and magnesium.

batholith: The largest intrusion size, greater than 100 square kilometers (38.5 square miles) in exposed surface area. Most are predominantly granitic rocks.

bedding or beds: The layers of sand, silt, carbonate, or other sediment that compose sedimentary rock.

bedrock: Solid rock that is part of the Earth's crust and is continuous beneath soils, lakes, and other unconsolidated surficial materials.

biotite: A dark brown, iron-rich mica—(K)(Fe, Mg) $_3$Al Si $_3$O $_{10}$(OH)—with such well-developed cleavage that it pulls apart easily into very thin sheets.

blueschist: A metamorphic rock containing minerals that can only develop in the very high pressure and relatively low temperatures of a subduction zone.

brachiopod: A bivalve mollusk, of the Phylum Brachiopoda. Brachiopods are similar to clams, but with vertically mirrored symmetry (right side=left side) rather than a clam's horizontal symmetry (top shell=bottom shell).

breccia: A rock composed of the broken fragments of other rocks, cemented together by finer material.

calcite: This soft mineral ($CaCO_3$), hardness of 3, is clear, white, or sometimes gray. It is a major constituent of limestones.

caldera: A flat-lying volcano, often unrecognized, that produces explosive eruptions of ash and tuff, followed by collapse of the volcano's remaining edifice into the evacuated magma chamber below.

carbonate: A rock, including limestone and marble, composed of carbonate minerals, especially calcite ($CaCO_3$), containing the carbonate radical, CO_3.

chert: A very hard chemically deposited sedimentary rock composed almost entirely of silica.

chlorite: A green metamorphic mineral—(Mg_5Al)($AlSi_3$) O_{10}(OH)$_8$—in the mica family. Chlorite provides the green hue to most metamorphosed greenstones and greenschists.

cinder cone: A steep-sided volcanic cone usually built of unconsolidated ejecta—scoria, bombs, or other basaltic particles—that erupt during gas-charged eruptions.

cirque: A circular or bowl-shaped "nest" carved by alpine glaciers at their mountainous point of origin.

clastic: A rock composed of broken pieces (clasts) or other rocks or minerals. Especially a sedimentary rock such as sandstone.

coal: The dark (usually black) highly compressed remains of woody plants.

columnar jointing: Columns, caused by multiple, equally spaced parallel cracks, which develop as igneous rocks cool and shrink in volume.

composite volcano: Also known as a stratovolcano, composite volcanoes build scenic, high cones. They produce a variety of products, including ash, tuff, basalt, andesite, and dacite lavas, lahars, and ignimbrites.

concretions: Rounded lumps of harder material found in sandstones and shales.

conglomerate: Sedimentary rocks consisting of rounded cobbles and pebbles in a sandy or muddy matrix. They look a bit like concrete, and the individual clasts may be 3 feet or more in diameter. Most conglomerates represent stream, river, or glacial deposits.

contact metamorphism: Changes in mineral content, composition, and/or texture of rock adjacent to an intrusion, produced by the heat and pressure of the intruding magma.

crossbedding: A bed in which subordinate layers have been deposited at an angle to the major bed itself by wind or water currents.

dacite: A silica-rich volcanic rock that is often glassy in appearance.

debris flow: Any flowing, moving mixture of water, rocks, sand, and clays, often including organic detritus including trees, stumps, and wood. Most often associated with volcanic activity.

dike: Narrow intrusions that crosscut the surrounding rock and serve as conduits for rising magma, often en route to an eruption.

diorite: A medium-gray, usually coarse-grained rock, intermediate between granitic rocks and gabbro.

dip: The angle that bedding, faults, or other rock structures makes with a horizontal plane. Expressed as the angular difference between the horizontal plane and the structure.

dolomite: Similar to limestone but composed of magnesium carbonate rather than calcium carbonate. Dolomite does not react with acid.

dome: The feature produced by the accumulation of viscous, silica-rich lavas into a huge, fractured mound directly above their vent. Domes are often a late phase of eruptions for stratovolcanoes, but may also erupt on their own. They are usually composed of rhyolite, obsidian, or dacite.

drumlin: An elongate hill composed of unsorted glacial gravels.

EarthScope Transportable Array (USArray): A 15-year program (2007–2022) sponsored by the National Science Foundation to place a dense network of permanent and portable seismographs in more than 2,000 locations across the continental United States. These seismographs record the seismic waves released by earthquakes that occur around the world. By analyzing the records of earthquakes obtained from this dense grid of seismometers, scientists can learn about Earth structure and dynamics and the physical processes controlling earthquakes and volcanoes. (See: http://www.usarray.org/researchers/obs/transportable.)

epicenter: The point on the ground surface that is directly (vertically) above the subsurface rupture (focus) of an earthquake.

epidote: A blocky, pistachio-green metamorphic mineral, $Ca_2Al_2(Fe^{3+};Al)(SiO_4)(Si_2O_7)O(OH)$.

erratic: A rock transported from one place and left in another, especially where the bedrock is dissimilar. Erratics are associated with Ice Age Floods as well as glaciers.

esker: A gravel-filled mold of a subsurface stream that ran beneath or within glacial ice. They are most common in continental glaciers, but may occur in very large alpine glacial systems.

exfoliation: A weathering process involving both chemical weathering of minerals and physical separation of rock, often owing to release of pressure as bedrock overburden is removed by erosion or uplift. The pattern produced by exfoliation is (usually) concentric layers, so that the "exfoliating" rock has the appearance of a layered onion.

extrusive: Igneous rocks, usually lava flows, which originate beneath the surface as fluid magma and erupt to solidify on the Earth's surface.

Farallon plate: An oceanic plate, which began subducting under the west coast of North America, then located in modern Idaho and Utah, during the Jurassic Period.

fault: A fracture or zone of fractured rock in Earth materials along which movement has occurred.

felsic: A light-colored igneous rock that contains more than 60 percent of light-colored minerals such as feldspar, quartz, and muscovite mica. Felsic also refers to the magmas from which these rocks crystallize.

fissure eruption: Eruptions of basaltic lavas from long cracks in the ground (fissures). Fissures often extend for miles.

flood basalts: Flood basalts, including the Columbia River basalts, mark regions where a mantle plume of basalt erupts. These eruptions flood thousands of square miles with fluid lava.

foliation: The planer structure that develops as a metamorphic rock undergoes heat and pressure. The intensity of pressure, and duration of metamorphic events, determine whether the rock's foliation is subtle or pronounced.

formation: Also known as a geological formation, this is a contiguous, and originally connected, sequence of strata or rocks that is deposited in similar conditions, usually over a relatively short time (geologically).

fumarole: A volcanic vent that releases steam and other gasses, often including sulfur.

gabbro: A mafic intrusive rock, dark green on freshly broken surfaces, and brown to dark green on weathered surfaces. Common minerals are pyroxene, plagioclase, and olivine. Gabbro has the same chemical and mineralogical compositions as basalt.

glacial polish: The polished surface crafted by ice and incorporated by sand moving over rock like fine emery cloth.

gneiss: This high-grade, metamorphic, banded rock is a sort of zebra-like granite. Minerals include quartz, feldspar, biotite, hornblende, and garnet; pyroxene and epidote may also be present.

graben: A down-faulted valley, or down-dropped block between two faults. Usually elongate. The up-throne blocks on either side are known as "horsts." Grabens usually develop where the crust is pulling apart or extending.

graded bedding: A bed of sedimentary rock with big particles at the bottom and progressively smaller ones on the top. Graded bedding develops when moving water first deposits large particles, and as the water slows, smaller particles settle on top of the larger sands or cobbles.

granite: A plutonic igneous rock containing about 45 percent quartz, 25 percent potassium (orthoclase) feldspar, and a variety of other minerals, especially muscovite, biotite, and hornblende. Because the Pacific Northwest's crust is low in potassium, true granite is uncommon here. Most Pacific Northwest rocks that look like granite are actually close cousins: granodiorite and tonalite.

granitic rock: A general term for intrusive igneous rocks that resemble granite. Generally light-colored, coarse-grained material composed of feldspar and quartz with biotite and hornblende as the dark minerals.

granodiorite: An intrusive igneous rock with greater than 20 percent quartz by volume where 65 to 90 percent of the feldspar is plagioclase. Mafic (dark, iron-rich) minerals are commonly biotite mica or hornblende.

greenschist: A low- to medium-grade metamorphic rock composed of much the same minerals as greenstone—chlorite and epidote—along with white (muscovite) mica. The term "greenschist" also applies to the metamorphic facies, or conditions, under which these rocks form.

greenstone: A low- to medium-grade metamorphosed volcanic rock rich in chlorite and/or epidote. The ancestor (or protolith) is most commonly basalt or andesite.

hanging valley: A tributary valley with a bottom or base notably higher than the stream it enters. There is usually a waterfall where the two valleys meet. Most commonly a feature of alpine glaciation, created when a tributary glacier feeds into the main valley glacier.

horn: A pointed spire produced when glaciers carve cirques into three or more sides of a peak.

hornblende: The jet-black color and usually elongate shape of this iron-rich mineral is distinctive. Hornblende is common in many andesites, but varies greatly in composition. Its typical formula is $(Ca,Na)_{2-3}(Mg,Fe,Al)_5(Al,Si)_8O_{22}(OH,F)_2$.

hornfels: Where a large granitic magma intrudes porous sandstone or shale, a specific type of contact metamorphic rock forms—a hornfels, which is usually a dark, dense, iron-rich, and often, mineralized, rock.

hotspot: A center of volcanic activity above a rising plume of magma that comes from a deep and stationary source in the Earth's mantle.

hydrothermal: Pertaining to hot waters usually related to intruding or cooling magmas. Many ore deposits are related to the circulation of hydrothermal fluids (hot waters and a variety of acids and other compounds) during the last stages of emplacement of a granite or granitic intrusion.

ignimbrite: A (usually) silica-rich volcanic rock produced by a pyroclastic flow or explosive eruption of hot, molten ash and gas. The word comes from the Greek *ignis* (fire) and *nimbus* (cloud). Ignimbrites are characterized by glassy, flattened, elongate clasts in the most densely welded center sections.

intrusion: A body of igneous rock of any size, formed beneath the surface, which forced its way into surrounding formations.

isotope: One of several atomic configurations of an element containing the same number of protons but varying numbers of neutrons.

joint: A fracture in the rock along which there is not displacement. Joints are usually related to the stresses experienced by a formation, or the cooling history of an igneous rock.

juvenile: Material, especially water, that originates from the Earth's interior.

dame: Small hills produced in glacial deposits as ice melts, depositing the gravels it was carrying on its surface, and thus producing an irregular set of hills.

dettle: A depression formed in glacial deposits by a leftover buried block of ice.

lahar: A flow of mud, rocks, and other detritus generated by melting glacial ice or snowpack, and sometimes generated or exacerbated by heavy rains. The name "lahar" is also applied to the resulting rock—a very coarse, unsorted conglomerate consisting of boulder-to-pebble-sized volcanic rocks (often from the volcano's summit area, and preponderantly dacite or andesite) in a fine-grained matrix of mud and ash.

lateral moraine: A long ridge of unsorted gravels carried along the edges of a glacier and deposited both as the glacier occupies the valley and as it melts. They are features associated only with alpine glaciers.

lava tubes; lava caves: A round "cave" left when hot lava simply runs out of its more solid conduit, much like floodwaters surging through a culvert and then subsiding. Pahoehoe lava flows often build a solid crust as the flow cools. However, fluid lava may continue to move beneath this crust, forming a closed, culvert-like conduit, known as a lava tube. When the eruption stops, the remaining fluid basalt flows out of the tube, leaving an open space.

limestone: A soft, chemical sedimentary rock, usually gray, composed mostly of calcium carbonate ($CaCO_3$). It falls into a broad classification of chemical sediments called "carbonates" that include dolomite (calcium-magnesium carbonate) and magnesite (magnesium carbonate).

lithophysae: Small, round cavities or areas of altered rock left by hot gas bubbles usually in silica-rich volcanic rocks (rhyolites and dacites). They may contain radiating crystals or concentric circles.

low-angle reverse fault: See **thrust fault**.

Ma: Standard geologic abbreviation for "mega annum," or a million years.

maar: A shallow, flat-floored crater produced when magma (usually basalt) encounters groundwater, a marsh, or a lake en route to the surface and its eruption may heat and boil the water, resulting in a violent steam explosion.

mafic: A dark igneous rock that contains a high percentage of iron, magnesium, and/or calcium. Mafic rocks usually contain the minerals olivine and pyroxene, along with a calcium-rich feldspar. Examples are basalt, anorthosite, and gabbro.

magma: Molten rock that occurs beneath the Earth's surface.

mantle: A major subdivision of Earth's internal structure. Located between the base of the crust and overlying the core, the mantle consists mostly of the dark, dense rock peridotite. It extends from about 25 miles in depth to 1,790 miles. The rocks of the mantle are very hot, and under intense pressure, and actually flow slowly.

marble: Metamorphosed limestone.

mélange: A chaotic zone or "broken" formation in which a variety of rocks of varying sizes, ages, and lithologies, from the mantle, deep seafloor, island arcs, and the continent, are incorporated into a fine-grained matrix.

metamorphic facies: A classification system that categorizes the conditions and products of metamorphism more precisely than metamorphic grade. Facies classification utilizes pressure and temperature conditions, as well as specific index minerals.

metamorphic rock: Metamorphic rocks (from the Greek term *met*—to change) are those that have been changed in a solid state by heat and/or pressure. They develop new minerals as well as distinctive new fabrics and texture.

migmatite: A distinctive, banded, coarse-grained metamorphic rock produced when subsurface rocks, usually granites, sandstones, or other silica-rich rocks, become hot enough to almost—but not quite—melt. Migmatite also contains very irregular lenses of granitic rock.

mineral: A naturally occurring, inorganic solid with a definite chemical composition and an ordered internal structure.

monocline: A sharp flexure or bend in rocks that does not change the overall orientation of the beds. It may just be a localized steepening of nearly flat sedimentary beds.

moraine: Derived from the old French *morre*, meaning muzzle or snout, this is a general term for any unsorted geologic debris (such as rocks, sand, gravels, and clay) carried by glacial ice and deposited at the sides or end of a glacier.

normal fault: Faults that accommodate crustal extension and have most motion in a near-vertical direction. One side is down-dropped relative to the other as the rock formations are tugged apart.

obsidian: A black, dark reddish brown, or (rarely) green, volcanic glass with the same silica-rich composition as rhyolite or dacite.

olivine: A mineral—$(Mg,Fe)_2SiO_4$—that is stable at anhydrous conditions and high temperatures. Fresh olivine rarely survives long at the surface; when it does, its transparent yellow-green color is distinctive.

ophiolite: A sequence of rocks that compose oceanic crust and upper mantle that has been uplifted and thrust onto the continent. From bottom to top an ophiolite consists of peridotite, gabbro, basalt, and sedimentary rocks, especially chert and fine-grained shale. Ophiolites may also include basaltic dikes, serpentinite (altered peridotite), and other rocks, including minor amounts of limestone. Ophiolites are large-scale features usually exposed over tens or hundreds of square miles.

outcrop: Bedrock exposed above the surface of the ground.

overturned-recumbent fold: In some folds, where force has been applied more from one direction than the other, the fold has been pushed over so it lies more or less on its side. This fold geometry is called an overturned fold (or, if the fold axis is horizontal, a recumbent fold.)

paholehoe: The ropy-textured variety of basaltic lava flow.

paleosol: An ancient soil (paleo-soil). In volcanic terrains, soils buried by hot lava flows may oxidize to a red color as hot steam interacts with iron in the soil (and may transport additional iron from the lava flow into the soil).

Pangaea: A supercontinent, consisting of all of Earth's present continents. Pangaea consisted of a southern megacontinent, Gondwana (Africa, South America, Australia, and Antarctica), and a northern megacontinent, Laurasia (North America, Europe, and Asia). Stretching virtually from the North to the South Poles, Pangaea collected about 300 million years ago and began to disperse about 200 million years ago.

pegmatite: A very coarse-grained granite that displays exceptionally large crystals at least 2 inches in length.

peridotite: A dark, dense rock, rich in magnesium and iron. Its principal minerals are pyroxene and olivine. Peridotite constitutes the Earth's mantle and is usually found on the surface as part of an ophiolite.

phenocryst: A mineral that you see easily with your naked eye. Phenocryst (feen' – O- crist) is a Greek term that means "visible crystals."

phyllite: A metamorphic rock intermediate between slate and schist. In this low-grade rock, clay minerals of the slate have started to transform into micas. In most phyllites, deformation helps to accentuate the foliation and flatten and crush existing minerals, giving a phyllite the appearance of very fine-grained schist.

pillow basalt, pillow lava: Basalt flows that are composed of rounded, globular, or tubular forms known as "pillows."

plagioclase feldspar: A common, light gray to clear mineral with a composition that varies systematically in calcium and sodium content, $NaAlSi_3O_8$ to $CaAl_2Si_2O_8$. It appears as light-colored, lath-shaped rectangles in many basalts, and forms the largest portion of most granitic rocks of the Wallowa, Wooley Creek, and Snoqualmie batholiths, and most other large intrusions in the Northwest.

platy jointing: Thin, usually horizontal cracks, which may create a layered appearance in volcanic rocks. This feature is especially common in andesites, but also affects basalt and even welded tuff. Volcanic rocks displaying platy jointing look more like shale—a completely unrelated sedimentary rock.

pluton: The term applies to any body of intrusive rock that formed and solidified beneath the Earth's surface, although it usually implies larger intrusions.

pumice: A highly vesicular, usually explosively erupted rhyolite or dacite.

pyrite: A brittle yellow, metallic mineral (FeS_2) often associated with gold and other ore deposits. Also known as fool's gold.

pyroclastic volcanic rocks: Rocks composed of particles of volcanic rocks and/or ash, usually produced and deposited by hot, explosive eruptions.

pyroxene: A dark green, stubby mineral with a highly variable composition. Most commonly: $Ca (Fe, Mg) (Si, Al)_2O_6$. Pyroxene is an essential ingredient of mafic and ultramafic rocks, basalt, gabbro, and peridotite.

quartz: A mineral (SiO_2) that is usually clear, smoky, or white. It is found in rhyolites as small, clear crystals, and granitic rocks, where it is the last mineral to solidify.

quartzite: An extremely hard, resistant metamorphic rock resulting from the application of heat and pressure to sandstone.

recessional moraine: As glaciers retreat, they may reach equilibrium, with their terminus at a new but higher elevation. Here, they build a recessional moraine—a sort of interim terminal moraine.

reverse fault: Reverse faults are caused by compression. As rocks are squeezed, one side is shoved up relative to the other as the rock formations first bend and then break. The slip-plane of a reverse fault is relatively steep (greater than 45 degrees); those with angles greater than 60 degrees are called "high-angle reverse faults" to emphasize this feature.

rhyodacite: A silica-rich volcanic rock, between dacite and rhyolite in composition. It is often produced by explosive eruptions, and may represent the composition of pumice or ash.

rhyolite: A light-colored, silica-rich volcanic igneous rock. It is often pink or reddish, the result of copious water thoroughly oxidizing their miniscule (1 to 3 percent) iron. Flow banding is a common feature. Rhyolite's name comes from the Greek word *rheos*, which means "flow."

sandstone: One of the best-known and most common sedimentary rocks. As its name implies, its principal ingredient is sand—meaning rock or mineral particles that are distinguishable by the unaided eye, and range from 0.08 to 0.0025 inches in diameter.

schist: A medium-grade highly foliated metamorphic rock whose principal minerals are muscovite (white) mica, biotite (black) mica, quartz, and feldspar. Schists have a bright luster imparted by abundant mica to reflect light.

scoria: A basalt lava that was forcefully mixed with steam and other volcanic gases and erupted explosively. Because scoria is crafted from an iron-rich basalt lava, it is darker than pumice, and because the iron oxidizes when exposed to air and water vapor, scoria is characteristically red or reddish brown.

seamount: A mountain, usually of volcanic origin and basaltic composition, that rises at least 1,000 meters (3,280 feet) from the seafloor.

serpentinite: A shiny green-to-black rock (occasionally white). Its name comes from the rock's shiny green appearance and seeming resemblance to the sheen of a snake's scales. It develops when peridotite is metamorphosed, usually by reacting with very hot (300 to 350 degree C, or 600 to 700 degree F) fluids (usually water) at depth of several to tens of miles beneath the surface.

shale: Usually a dark-colored, fine-grained, dull-appearing, and very soft sedimentary rock.

shield volcano: Low, mound-shaped volcanoes, composed almost exclusively of basalt, with unimpressive height but deceptively huge volume.

sill: A thin intrusion parallel to surrounding strata.

slate: A relatively dense, low-grade metamorphic rock. In the Northwest, most slates are dark gray, but some are more varied, from reddish brown to dark green. Slate is a low-grade rock that develops from shale at relatively low temperatures (300 degrees C, or 570 degrees F) and pressures.

spheroidal weathering: Creates rounded forms as rocks weather. The spherical form develops as organic acids seep into cracks in fresh rock, softening the corners. Over time, the sharp corners are transformed into clays by reaction with these weak acids. The remaining solid rock retreats, eventually leaving rounded forms instead of square.

striations: Grooves and scratches on bedrock. Most commonly, the rocks carried in the moving glacial ice produce striations. They indicate the direction of glacial ice, and to some degree, its intensity.

strike-slip fault: Faults that transfer rocks laterally, but not up or down.

stromatolite: Calcareous algae and/or cyanobacteria that build photosynthetic mound-shaped structures.

subduction zone: An area at a convergent plate boundary where an oceanic plate is descending into (subducting) the mantle beneath another plate.

syenite: Pronounced cy'-an-ite, this granite-like intrusive rock contains mostly feldspar and virtually no quartz. Like granite, it is extremely rare in the Northwest.

syncline: A "down-fold" whose limbs continue upward away from the fold axis.

terminal moraine: The pile of unsorted gravel left at the snout or end (terminus) of a glacier. This material is carried in, on, or beneath the glacier, and deposited when the ice melts.

thrust fault: Also called a "low-angle reverse fault," a thrust fault is driven by compression. Their fault planes have a much lower angle (45 degrees or less) than reverse faults, but the direction of motion is the same—one side pushed up relative to the other.

tonalite: A granitic intrusive igneous rock, abundant in the Northwest, which looks similar to granite. It consists of less than 10 percent potassium-bearing feldspars. Biotite and/or hornblende are common minerals. It is one of the most abundant granitic rocks in the Northwest.

trace fossils: The tracks of marine worms, or their soft-bodied remains, which occur as puzzling scribbles in Pacific Northwest marine sedimentary rocks.

tree molds: Also known as a "lava cast," a tree mold is the hollow space left when a lava flow envelops a tree, solidifies around the flaming trunk, and ultimately, the tree burns to charcoal.

tuff: A soft rock composed of volcanic ash. Most tuffs are "air fall"—material that was blasted into the air and then fell to the ground as a cool solid.

turbidites: A sequence of alternating sandstones and shales that accumulates to great thickness in deep-sea fans and other deep-water settings.

varve: An annual layer of sediment, usually deposited in a lake and often associated with glacial conditions. Dark layers are usually deposited in summer, when organic matter is available, and lighter layers are deposited in winter.

veins: Streaks of usually white minerals that crosscut the overall patterns in the rock. Vein are produced by the release and migration of fluids through cracks and other surfaces within rocks, a process that often accompanies metamorphism.

vent: The location of a volcanic eruption, or the orifice from which volcanic materials erupt.

vesicles: Holes created in the molten lava by volcanic gas.

volcanic bomb: During its few fleeting moments of flight, a chunk of molten basalt lava flung into the air by an eruption may mold itself into an aerodynamic shape. These graceful, teardrop forms are known as volcanic bombs.

welded tuff: Formed when explosive eruptions eject hot clouds of ash and gas that travel rapidly and come to rest while the ash particles are still hot and semi-molten. Hot particles of pumice and ash adhere or "weld" together.

xenoliths: Clots of rock from another source that are found in intrusive or extrusive igneous rocks. Most commonly, these "foreign rocks" are found as darker clots in granitic rock.

zircon: A very stable mineral $(ZrSiO_4)$ that usually occurs as very small crystals in granitic rocks. Zircon contains a relatively high concentration of uranium, and hence is useful for radiometric dating. Because it is resistant to chemical and physical weathering, it withstands repeated episodes of sedimentation, exhumation, and weathering.

About the Author

ELLEN MORRIS BISHOP has documented, researched, and photographed the geology of the Pacific Northwest, with special interest in exotic terranes. She holds a PhD in geology from Oregon State University. Her work has won the Frances Fuller Victor Oregon Book Award for nonfiction and the Oregon State University Distinguished Alumni Award.

Ellen Morris Bishop. Preferred habitat: With camera, Lakes Basin, Wallowa Mountains.

Meesha: Golden Retriever/ German Shepherd (?) mix. Adopted from Oregon Humane Society. Certified therapy dog.

Dundee: Australian Shepherd, Adopted from Oregon Humane Society.

Kyla: Border Collie. Adopted from Oregon Border Collie Rescue.

Diesel: Australian Shepherd/ Border Collie. Adopted from Montana family.